GREEN HETEROGENEOUS WIRELESS NETWORKS

GREEN HETEROGENEOUS WIRELESS NETWORKS

Muhammad Ismail

Texas A&M University at Qatar, Doha, Qatar

Muhammad Zeeshan Shakir

University of the West of Scotland, Glasgow, UK

Khalid A. Qaraqe

Texas A&M University at Qatar, Doha, Qatar

Erchin Serpedin

Texas A&M University, College Station, Texas, USA

Library of Congress Cataloging-in-Publication Data

Names: Ismail, Muhammad, 1985 November 20- author. | Shakir, Muhammad
  Zeeshan, author. | Qaraqe, Khalid A., author. | Serpedin, Erchin, 1967-
  author.
Title: Green heterogeneous wireless networks / Muhammad Ismail, Muhammad
  Zeeshan Shakir, Khalid A. Qaraqe, Erchin Serpedin.
Description: Chichester, UK ; Hoboken, NJ : John Wiley & Sons, 2016. |
  Includes bibliographical references and index.
Identifiers: LCCN 2016010885 (print) | LCCN 2016015082 (ebook) | ISBN
  9781119088059 (cloth) | ISBN 9781119088028 (pdf) | ISBN 9781119088035
  (epub)
Subjects: Green communications | Energy efficiency | Heterogeneous wireless medium | multi-homing.
Classification: LCC TK5105.78 .I86 2016 (print) | LCC TK5105.78 (ebook) | DDC
  004.6/8–dc23
LC record available at http://lccn.loc.gov/2016010885

A catalogue record for this book is available from the British Library.

ISBN: 9781119088059

Typeset in 10/12pt TimesLTStd by SPi Global, Chennai, India
Printed and bound in Malaysia by Vivar Printing Sdn Bhd

1   2016

Contents

Preface

This book focuses on the emerging research topic 'green (energy- efficient) wireless networks' that has drawn huge attention recently from both academia and industry. This topic is highly motivated due to important environmental, financial and quality-of-experience (QoE) considerations. Due to such concerns, various solutions have been proposed to enable efficient energy usage in wireless networks, and these approaches are referred to as green wireless communications and networking. The term 'green' emphasizes the environmental dimension of the proposed solutions. Hence, it is not sufficient to present a cost-effective solution unless it is eco-friendly.

In this book, we mainly focus on energy-efficient techniques in base stations (BSs) and mobile terminals (MTs) as they constitute the major sources of energy consumption in wireless access networks, from the operator and user perspectives. Furthermore, this book targets the heterogeneous nature of the wireless communication medium, and therefore, the book is entitled 'Green Heterogeneous Wireless Networks'. The wireless communication medium has become a heterogeneous environment with overlapped coverage due to the co-existence of different cells (macro, micro, pico and femto), networks (cellular networks, wireless local areas networks and wireless metropolitan area networks) and technologies (radio frequency, device-to-device (D2D) and visible light communications (VLC)). In such a networking environment, MTs are equipped with multiple radio interfaces. Through multi-homing capability, an MT can maintain multiple simultaneous associations with different networks. Besides enhancing the achieved data rate through bandwidth aggregation, the heterogeneous wireless medium together with the multi-homing service can enhance the energy efficiency of network operators and mobile users.

This book consists of three parts. The first part provides an introduction to the 'green networks' concept and identifies the key problems associated with the existing green solutions. The first part consists of two chapters. The first chapter discusses the need for green (energy-efficient) communications, the modelling techniques used for energy efficiency and call traffic in wireless networks and different performance metrics. The second chapter reviews the existing solutions for green networking at different call traffic load conditions. It covers the green solutions adopted by different standards (e.g. 3GPP). Limitations and key problems of the existing solutions are also discussed.

The second part of the book targets the green multi-homing resource allocation problem, and it consists of four chapters. The first chapter introduces the green multi-homing resource allocation problem and discusses its potential benefits and challenges. The limitations of the existing multi-homing green solutions are discussed and practical aspects that should be accounted for are presented to assist engineers and network operators in building green multi-homing solutions. These limitations and practical considerations are then discussed in detail in the following chapters. The second chapter addresses a major limitation of practical value in the existing green downlink multi-homing resource allocation strategies. Specifically, the existing solutions implicitly assume that all networks are willing to cooperate unconditionally for energy saving, which is not practical, and therefore, we present a novel win–win resource allocation mechanism that enables energy saving for network operators. Furthermore, a radio resource allocation framework that accounts for the in-device coexistence (IDC) interference between the LTE and WiFi networks is also presented. The third chapter addresses a major limitation of existing research on green uplink multi-homing resource allocation for data calls. Specifically, existing solutions adopt a single-user system model, which is not practical for uplink resource allocation, and hence we present a novel joint bandwidth and power allocation framework in a multi-user system that maximizes the minimum energy efficiency among all MTs in service. In addition, uplink resource allocation for sustainable multi-homing video transmission is also discussed. An energy management subsystem that adapts the MT energy consumption during the call to achieve at least the target video quality lower bound is presented. The last chapter of the second part of the book presents a novel framework that integrates femto cells with VLC for a green downlink multi-homing resource allocation strategy to exploit jointly their benefits in energy saving while overcoming their practical limitations in terms of VLC reliability and femto-cell high energy consumption as compared to VLC.

The third part of the book addresses green network management solutions and consists of four chapters. The first chapter addresses BS on–off switching methods for energy saving. Two mechanisms are presented to serve the mobile users while switching off the BSs. One mechanism relies on a dense deployment of small-cells while the other mechanism relies on a cooperative networking technique. Furthermore, existing solutions mainly shift the energy consumption burden from the network operators to mobile users, which is not practical as it will drain the mobile user terminals at faster rates, and consequently, we present a novel dynamic planning approach with a balanced energy saving strategy for network operators and mobile users. The second chapter presents a novel deployment model for small-cells and cell-on-edge deployment to enhance energy efficiency of the networks. The third chapter presents a novel deployment of D2D communications and their successful integration into heterogeneous networks. This chapter presents also an end-to-end analysis of power consumption in the whole network and stresses out the significance of device-centric communications for 'greener networks'. The last chapter in this part of the book presents an emerging device centric green approach for content exchange/download between the devices by exploiting the multihoming and packet split over multiple interfaces in D2D links.

MUHAMMAD ISMAIL, MUHAMMAD ZEESHAN SHAKIR,
KHALID A. QARAQE, and ERCHIN SERPEDIN
Doha, Qatar
June 2016.

Acknowledgements

The authors would like to acknowledge their research collaborators for the joint research effort on many topics of mutual interest that helped the realization of this book. Special thanks go to Drs Weihua Zhuang, Mohamed Kashef, Mohamed Abdallah, Mohamed Marzban, Mohamed Khairy, Amila Gamage, Sherman Shen, Hafiz Yasar Lateef, Amr Mohamed, Mohamed-Slim Alouini, Hina Tabassum and Muhammad Ali Imran.

The authors would like to acknowledge the support from Qatar National Research Fund offered through NPRP.

Dedication

Muhammad Ismail dedicates this book to his beloved wife Noha, lovely sister Dina and dear parents Ismail and Wafaa.

Muhammad Zeeshan Shakir dedicates this book to his family members and friends for their support.

Khalid A. Qaraqe dedicates this book to the memory of his amazing father Ali, beloved wife May and lovely family.

Erchin Serpedin thanks his family and collaborators for their support.

Part One

Introduction to Green Networks

1

Green Network Fundamentals

Efficient energy usage in wireless networks has drawn significant attention from both academia and industry, mainly because of critical environmental, financial, and quality-of-experience (QoE) concerns. Research efforts have led to various solutions that allow efficient use of energy in wireless networks. Such approaches are referred to as *green wireless communication and networking*. Throughout this book, our main focus is on developing energy-efficient communication techniques in base stations (BSs) and mobile terminals (MTs), as they represent the major sources of energy consumption in wireless access networks, from the operator and user perspectives, respectively, while accounting for the heterogeneous nature of the wireless communication medium. Towards this end, the first two chapters of the first part of this book are dedicated to introducing the background concepts of green networking. The first chapter discusses the need for green (energy-efficient) communications, the modelling techniques used for energy efficiency and call traffic in wireless networks, and different conflicting performance metrics. Building on such a background, the second chapter reviews the state-of-the-art green communication solutions and analytical models proposed for network operators and mobile users at different traffic load conditions, and points out their major shortcomings.

1.1 Introduction: Need for Green Networks

In response to the increasing demand for wireless communication services during the past decade, there has been wide deployment of wireless access networks [1]. By definition, a wireless access network is a wireless system that uses BSs and access points (APs) to interface MTs with the core network or the Internet [2]. Hence, the main components of a wireless access network are BSs/APs and MTs [3]. BSs/APs are mainly in charge of radio resource control and user mobility management, and provide access to the Internet. MTs are equipped with processing and display capabilities, and provide voice services, video streaming, and data applications to mobile users. Currently, MTs are provided with multiple radio interfaces, and mobile users can connect to different networks, such as cellular networks, wireless local area networks (WLANs), and wireless metropolitan area networks (WMANs), and enjoy single-network and/or multi-homing services [4–6].

Green Heterogeneous Wireless Networks, First Edition. Muhammad Ismail, Muhammad Zeeshan Shakir, Khalid A. Qaraqe and Erchin Serpedin.
© 2016 John Wiley & Sons, Ltd. Published 2016 by John Wiley & Sons, Ltd.

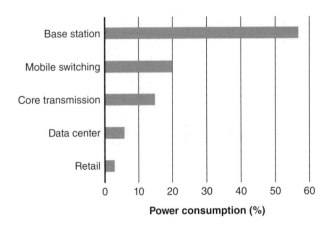

Figure 1.1   Breakdown of power consumption of a wireless cellular network [7]

From the network operator side, BS is the main source of energy consumption in the wireless access network [2]. The breakdown of a cellular network's typical power consumption is shown in Figure 1.1, which shows that almost 57% of the operator's total power consumption is in the BS [2, 8, 9]. Worldwide, there are about 3 million BSs, which consume in total 4.5 GW of power [10]. From the user side, it has been estimated that there exist roughly 3 billion MTs in the world with a total power consumption of 0.2–0.4 GW [11]. Such high energy consumption of wireless access networks has triggered environmental, financial, and QoE concerns for both network operators and mobile users.

From an environmental standpoint, the telecommunications industry is responsible for 2% of the total CO_2 emissions worldwide, and this percentage is expected to double by 2020 [12]. As shown in Figure 1.2, the mobile communications sector has contributed 43% of the telecommunication carbon footprint in 2002, and this contribution is expected to grow to 51% by 2020 [14]. Furthermore, the MT rechargeable batteries' expected lifetime is about 2–3 years and manifests in 25,000 t of disposed batteries annually, a factor that raises environmental concerns (and financial considerations for the mobile users as well) [15]. In addition, the high energy consumption of BSs and MTs is a source of high heat dissipation and electronic pollution [16]. From a financial standpoint, a significant portion of a service provider's annual operating expenses is attributed to energy costs [17, 18]. Technical reports have indicated that the cost of energy bills of service providers ranges from 18% (in mature markets in Europe) to 32% (in India) of the operational expenditure (OPEX) [19, 20]. The energy expenses reach up to 50% of the OPEX for cellular networks outside the power grid [21, 22]. Finally, from a user QoE standpoint, it has been reported that more than 60% of mobile users complain about their limited battery capacity [23]. In addition, the gap between the MT's offered battery capacity and the mobile users' demand for energy is growing exponentially with time [24]. Consequently, the MT's operational time between battery chargings has become a crucial factor in the mobile user's perceived quality-of-service (QoS) [25].

The aforementioned concerns have triggered increasing demand for energy-efficient solutions in wireless access networks. Research efforts carried out in this direction are referred to as *green network solutions*. The term 'green' confirms the environmental dimension of the proposed approaches. Therefore, a cost-effective solution that is not eco-friendly is

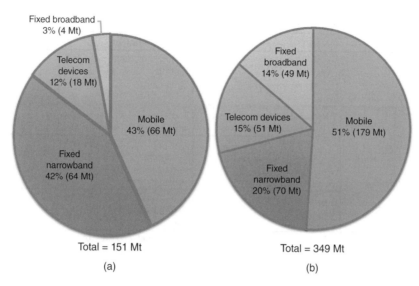

Total = 151 Mt

(a)

Total = 349 Mt

(b)

Figure 1.2 Carbon footprint contribution by the telecommunications industry: (a) 2002 and (b) 2020 [13]

not attractive. For instance, having a cost-effective electricity demand schedule for a network operator that relies on different electricity retailers, in a liberated electricity market, is not considered a green solution if it does not ensure that the proposed solution is also eco-friendly in terms of the associated carbon footprint [26]. The objectives of the green wireless communications and networking paradigm are, therefore, (i) reducing energy consumption of communication devices and (ii) taking into account the environmental impacts of the proposed solutions.

In order to develop/analyse a green networking solution, an appropriate definition of energy efficiency/consumption for network operators and mobile users should be formulated. This definition should account for the power consumption, throughput, traffic load models, and conflicting performance metrics for network operators and mobile users. The first chapter of this book is dedicated to building this necessary background.

1.2   Traffic Models

Some energy-efficiency and consumption models are defined on the basis of the temporal fluctuations in the traffic load. In addition, different green approaches can be adopted at different traffic load conditions. Furthermore, some green approaches rely on the temporal and spatial fluctuations in the traffic load to save energy. For instance, in order to determine the sleep duration of a BS or MT, traffic models are used to probabilistically predict the idle period duration, as will be presented in Chapter 2. Moreover, the performance evaluation of the green approaches should be carried out using an appropriate traffic model. Consequently, it is necessary to gain a better understanding of the different traffic load models proposed in the literature before introducing energy efficiency and consumption models as well as green solutions.

Table 1.1 Summary of different traffic models [27]

Model					Comments	References
Static					It does not capture the MT mobility and the traffic dynamics	[23, 28–34]
Dynamic	Spatial	Regional traffic load density			It defines a location-based traffic load density	[35]
		Stochastic geometry			BSs and MTs are located according to a homogeneous Poisson point process	[18]
		FSMC			It models the spatial distribution of MTs within a cell	[36]
	Temporal	Long-scale			The model captures traffic fluctuations over the days of the week	[17, 37–39]
		Short scale	Flow-level	Poisson-exponential	It models call arrivals as a Poisson process and call departures as an exponential distribution	[12, 40–42]
				FSMC	The number of calls within a cell is represented by a state in a Markov chain	[36]
			Packet-level	Infinite buffer	It models the number of backlogged packets in an MT buffer with infinite capacity	[43]
				Finite buffer	It models the number of backlogged packets in an MT buffer with finite capacity	[44]

Overall, the traffic modelling can be categorized into two classes, as shown in Table 1.1. The first class is referred to as the *static model* and assumes a fixed set of MTs, $\mathcal{M}$, that communicate with a fixed set of BSs, $\mathcal{S}$ [23, 28–34, 45]. The static model suffers from several limitations. First, it does not consider the mobility of MTs in terms of their arrivals and departures. Second, it does not capture the call-level or packet-level dynamics in terms of call duration, packet arrival, and so on. On the other side, the second class, which is referred to as the *dynamic model*, captures the spatial and temporal fluctuations of the traffic load, and is discussed next in detail.

1.2.1 Traffic Spatial Fluctuation Modelling

Studies have indicated that traffic is quite diverse even among closely located BSs, as shown in Figure 1.3, [37, 38]. As a result, different models have been proposed in the literature to reflect the spatial fluctuations in call traffic load [18, 35, 36].

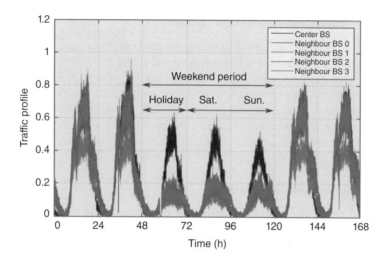

Figure 1.3 Spatial and temporal traffic fluctuations [38]

Location-based traffic load density is one approach to capture traffic spatial fluctuations [35]. In this context, a geographical region is covered by a set $\mathcal{S}$ of BSs and the region is partitioned into a set of locations. In a given location x, the file transfer request arrivals follow an inhomogeneous Poisson point process (PPP) with an arrival rate $\lambda(x)$ per unit area. The file sizes are independently distributed with mean $1/\mu(x)$ at the location. Consequently, the traffic load density is given by $\varrho(x) = \lambda(x)/\mu(x) < \infty$, which is used as a measure of the spatial traffic variability.

The aforementioned approach adopts a pre-defined set of BSs, $\mathcal{S}$, with specific locations. An alternative approach, which is more suitable for a design stage, defines the locations of BSs based on the stochastic geometry theory [18]. Hence, the network's n BS locations follow a homogeneous PPP, Θ_n, with intensity θ_n in the Euclidean plane. Similarly, MTs are located according to a different independent stationary point process with intensity θ_m. According to the stationary PPP Θ_n, the distance between an MT and its serving BS, D_m, follows the same distribution regardless of the MT's exact location. The probability density function (PDF) of D_m is expressed as [18]

$$f_{D_m}(d) = 2\pi\theta_n d \exp(-\theta_n \pi d^2), \quad d > 0. \tag{1.1}$$

The aforementioned models reflect the spatial variability of the traffic among different cells. To capture the spatial distribution variability of MTs within a given cell i, a finite-state Markov chain (FSMC) model is adopted [36]. This model classifies the MTs into G groups according to cell i's radius. Assuming there are M MTs in cell i, a G^M spatial location distribution is considered within the cell. Thus, the FSMC model presents $\mathcal{L} = \{L_1, \ldots, L_{G^M}\}$ states. The state transition probability $\Pr\{L_i(t+1) = v_i | L_i(t) = u_i\}$ is the probability of the spatial distribution of the MTs within the cell i at time slot $t+1$ to assume v_i, given that it was u_i at time slot t, where $u_i = \{u_{i,1}, \ldots, u_{i,M}\}$ and $v_i = \{v_{i,1}, \ldots, v_{i,M}\}$. Following this model, the dynamic fluctuations in the number of MTs in different regions within the cell can be captured.

1.2.2   Traffic Temporal Fluctuation Modelling

Two different time scales can capture the temporal fluctuations in the traffic load [12, 39]. The first time scale is a long-term one that reflects the traffic variations over the days of the week. Such a model can help in evaluating different energy-efficient approaches for network operators, as it captures both high and low call traffic load conditions. The second time scale is a short-term one that reflects the call (packet) arrivals and departures of the MTs. Such a model plays a vital role in evaluating energy-efficient resource allocation schemes for MTs and BSs. In the following subsections, we describe the two scales.

1.2.2.1   Long-Term Traffic Fluctuations

Real call traffic traces demonstrate a sinusoidal traffic profile in each cell, as shown in Figure 1.3, [17, 38]. During daytime (11 am–9 pm), traffic is much higher than that during nighttime (10 pm–9 am) [17, 37]. Furthermore, during weekends and holidays, the traffic profile, even during the peak hours, is much lower than that of a normal week day [17]. The traffic profile during a weekday is 10% less than its peak value 30% of the time, and this increases to 43% of the time during weekends [17]. This behaviour can be captured using an activity parameter $\psi(t)$, which specifies the percentage of active subscribers over time t, as shown in Figure 1.4 [39]. Denote p as the population density of users per km^2, N as the number of operators (each being able to carry $1/N$ of the total traffic volume), and M_k as the fraction of subscribers with an average data rate r_k for terminal type k (e.g. smart phone and tablet). Hence, the traffic demand, in bits per second per km^2, is given by

$$A(t) = \frac{p}{N}\psi(t)\sum_{k} M_k r_k. \tag{1.2}$$

Studies have indicated that the traffic load difference between two consecutive days for 70% of the BSs is less than 20% [37]. As a result, the long-term fluctuations in call traffic load can

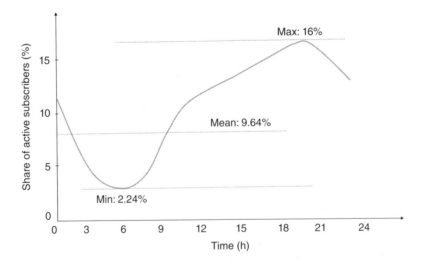

Figure 1.4   Average daily data traffic profile in a European country [39]

be estimated from the historical mobile traffic records; that is, the activity parameter $\psi(t)$ and the average data rate r_k can be inferred in practice from historical data.

1.2.2.2 Short-Term Traffic Fluctuations

Two categories can be distinguished for short-term traffic fluctuation models, namely call (flow)-level and packet-level models. Call (flow)-level models are useful in designing and evaluating green resource scheduling mechanisms at both BSs and MTs under high call traffic load. For myopic resource allocation solutions, the call arrivals are modelled using a Poisson process with rate λ, and the call durations are represented by an exponential distribution [12, 40–42]. Dynamic resource allocation solutions rely on FSMC to model traffic dynamics in terms of call arrivals and departures [36]. In this model, the number of calls in a given cell i is captured by an M-state Markov chain, with the state set $\mathcal{M} = \{0, 1, \ldots, M-1\}$. The state transition probability $\Pr\{M_i(t+1) = m_i | M_i(t) = \widetilde{m}_i\}$ is the probability of having m_i MTs within cell i at time slot $t+1$, given that there were $\widetilde{m}_i$ MTs at time slot t, where m_i, $\widetilde{m}_i \in \mathcal{M}$.

In a low call traffic load condition, packet-level traffic models are useful in designing and evaluating green resource solutions (on–off switching) at the BSs and MTs, through modelling the BS/MT buffer dynamics in terms of packet arrival and transmission [43, 44]. For an infinite buffer size, the MT buffer dynamics can be expressed as

$$o_m(t+1) = \max\{o_m(t) + a_m(t+1) - z_m(t), a_m(t+1)\}, \tag{1.3}$$

where $o_m(t)$, $a_m(t)$, and $z_m(t)$ are the numbers of backlogged packets in the buffer, arriving packets, and transmitted packets, for MT m in time slot t, respectively. For a buffer with a finite size F, the MT buffer dynamics can be represented by

$$o_m(t+1) = \min\{o_m(t) + a_m(t+1) - z_m(t), F\}. \tag{1.4}$$

The models (1.3) and (1.4) are used to investigate the optimal on–off switching mechanisms for the radio interfaces of MTs to achieve energy-efficient (green) communications at a low call traffic load condition, a topic that will be addressed in Chapter 2.

1.3 Energy Efficiency and Consumption Models in Wireless Networks

Following the temporal and spatial fluctuations in traffic load, this section summarizes different definitions that have been proposed in the literature to assess energy consumption/efficiency of wireless networks. Towards this end, we first present different throughput and power consumption models for BSs and MTs.

1.3.1 Throughput Models

The utility obtained from the wireless network in exchange for its consumed power is expressed most of the time in terms of the achieved throughput. In this context, we first introduce the concepts of aggregate BS capacity C_s, area spectral efficiency T_s, and user-achieved data rate R_m, which will be used in the energy efficiency definitions to be presented later.

1.3.1.1 Network Side

The BS aggregate capacity C_s for BS s is measured using Shannon's formula as follows [26]

$$C_s = B_s \log_2 \det(I + PH), \tag{1.5}$$

where B_s denotes the total bandwidth of BS s, I represents the unit matrix, P is the transmission power vector of BS s to every MT m in service, and H stands for the channel gain matrix between BS s and each MT m, which accounts for the channel's fast fading, noise, and interference affecting the radio transmission. The BS capacity C_s in (1.5) is measured in bits per second (bps).

At a low call traffic load condition, the area spectral efficiency T_s provides a better representation of the BS's attained utility than the BS's aggregate capacity since it accounts for the coverage probability, which matters the most at such a condition [18]. Specifically, T_s measures the BS throughput while considering the coverage probability. Denote $\Pr\{\gamma_{x \to u} > \zeta\}$ as the success probability of the signal-to-noise ratio (SNR) γ received by an MT at location u from a given BS at some location x satisfying a certain QoS threshold ζ. Averaging the success probability $\Pr\{\gamma_{x \to u} > \zeta\}$ over the propagation range to location u yields the coverage probability $\mathbb{P}_s(\zeta)$. For BS s, the area spectral efficiency T_s measured over a unit area is expressed as

$$T_s = \mathbb{P}_s(\zeta) \log_2(1 + \zeta). \tag{1.6}$$

1.3.1.2 Mobile Terminal Side

While the definitions in (1.5) and (1.6) are mainly from the operator side, two definitions can be used to quantify the mobile user's attained utility (in terms of the achieved data rate R_m in the uplink by MT m) in exchange for the MT power consumption. Given the instantaneous channel state information (CSI), the achieved data rate R_m in bps can be expressed as [16, 28, 45, 46]

$$R_m = B_m \log_2\left(1 + \frac{\gamma_m}{\Gamma}\right), \tag{1.7}$$

where B_m stands for the uplink allocated bandwidth to MT m, γ_m represents the SNR of MT m received at the destination, and Γ denotes the SNR gap between the channel capacity and a practical coding and modulation scheme. For the Shannon formula, $\Gamma = 1$. Reporting instantaneous CSI from each MT to the serving BS, in order to determine (1.7), leads to a large signalling overhead. In order to reduce the associated signalling overhead, a statistical CSI is used. Consequently, R_m in bps is expressed as

$$R_m = \mathbb{E}_H\left[B_m \log_2\left(1 + \frac{\gamma_m}{\Gamma}\right)\right], \tag{1.8}$$

where $\mathbb{E}_H$ represents the expectation over the channel state H.

1.3.2 Power Consumption Models

In order to attain the aforementioned utilities in (1.5)–(1.8), power is consumed at both the network side and user side. In the literature, different models are proposed to capture such a power consumption, as summarized in Table 1.2. These models are next discussed.

Table 1.2 Summary of different power models proposed in the literature [27]

	Model			Comments	References
BS	Operation only	Large-cell	Ideal	The BS consumes no power when idle, that is, the BS consists only of energy proportional devices	[35]
			Realistic	The model captures the BS traffic load independent power consumption	[12, 18, 21, 26, 35, 39]
		Femto-cell	Load independent	The BS power consumption does not depend on the offered traffic load	[47]
			Load dependent	The BS power consumption relies on traffic load, packet size, and has an idle part	[48]
		Including temporal fluctuations		The model accounts for full load, half load, and idle traffic conditions	[7]
		Backhaul power consumption		The model defines power consumption for micro-wave and optical fiber backhaul links	[49]
	Operation and embodied			Besides the operation power, it accounts for the consumed energy in BS manufacturing and maintenance	[19]
MT	Transmission power only	Without power amplifier efficiency		The model does not account for the transmitter power amplifier efficiency	[29, 46, 50]
		With power amplifier efficiency		The model accounts for the transmitter power amplifier efficiency	[16, 30, 45, 51]
	Including circuit power	Constant		The circuit power consumption is given by a constant term independent of the bandwidth and data rate	[16, 23, 28, 31, 45, 52]
		Bandwidth scale		The circuit power consumption scales with the MT assigned bandwidth	[53]
		Data rate scale		The circuit power consumption scales with the MT achieved data rate	[51]
	Including reception power			Besides the transmitter and circuit power consumption, the model also accounts for the receiver power consumption	[16,53]

1.3.2.1 Network Side

The total power consumption P_n of a wireless access network n, from the network operator perspective, can be captured using the aggregate power consumption of the network BSs. Recently, in addition to the BS power consumption, more emphasis is put on the

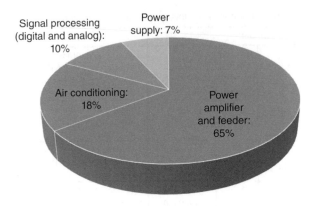

Figure 1.5 Percentage of power consumption at different components of a large-cell BS [27]

Table 1.3 Power consumption profile for a femto-cell BS [27]

Hardware component	Power consumption (W)	Percentage (%)
Microprocessor	1.7	
Associated memory	0.5	26.4
Backhaul circuitry	0.5	
FPGA	2	
Associated memory	0.5	39.2
Other hardware functions	1.5	
RF transmitter	1	
RF receiver	0.5	34.3
RF power amplifier	2	

backhaul power consumption, due to the information exchange among BSs for cooperative transmission/networking. Next, we will outline the different power consumption models proposed for BSs and backhauls.

For a large-cell BS (macro- and micro-BS), Figure 1.5 illustrates the power consumption percentage of different components of the BS. Furthermore, the power consumption profile of a femto-cell BS is shown in Table 1.3. According to Figure 1.5 and Table 1.3, the following facts turn out:

- The signal processing part is responsible for most of the power consumption in a femto-cell BS as opposed to a large-cell BS (namely, 65.6% and 10% for femto and large-cell BSs, respectively).
- The radio frequency (RF) transmission/reception power consumption in a femto-cell BS is almost half of that of a large-cell BS, with only 19.6% of the power consumed in the femto-cell BS power amplifier as opposed to 65% in a large-cell BS.

In the literature, different models are adopted to represent the BS power consumption P_s. For a large-cell BS, the simplest model is an ideal load-dependent representation, which assumes that the BS consumes no power in its idle state, that is, the BS consists of energy-proportional devices [35]. Hence, the BS power consumption can be expressed as

$$P_s = \rho P_{ts}, \tag{1.9}$$

where ρ stands for the system traffic load density, and P_{ts} denotes the BS's transmitted power. The major limitation with such a model is that it is unrealistic, as the power consumption of some BS components in reality is not load-dependent, as shown in Figure 1.5 (e.g. power supply and air conditioning). To capture the power consumption of both load-dependent and load-independent components in the BS, a more sophisticated model assumes the following expression [39]

$$P_s = \frac{\dfrac{P_{ts}}{\xi(1 - \sigma_{\text{feed}})} + P_{\text{RF}} + P_{\text{BB}}}{(1 - \sigma_{\text{DC}})(1 - \sigma_{\text{MS}})(1 - \sigma_{\text{cool}})}, \tag{1.10}$$

where P_{RF} represents the RF power consumption, P_{BB} denotes the baseband unit power consumption, ξ is the power amplifier efficiency, and σ_{feed}, σ_{DC}, σ_{MS}, and σ_{cool} stand for the losses incurred by the antenna feeder, DC–DC power supply, main supply, and active cooling, respectively. The model (1.10) is further approximated using a linear (affine) function for simplicity [12, 18, 21, 26, 35]. The affine function consists of two components to represent P_s. The first term is denoted by P_f and represents a fixed (load-independent) power component that captures the power consumption at the power supply, cooling, and other circuits. The second term is a load-dependent component. The affine model is expressed as

$$P_s = \Delta_s P_{ts} + P_f, \tag{1.11}$$

where Δ_s is the slope of the load-dependent power consumption.

For a femto-cell BS, the power consumption model is described by Deruyck et al. [47]

$$P_s = P_{\text{mp}} + P_{\text{FPGA}} + P_{\text{tx}} + P_{\text{amp}}, \tag{1.12}$$

where P_{mp}, P_{FPGA}, P_{tx}, and P_{amp} denote the power consumption of the microprocessor, field-programmable gate array (FPGA), transmitter, and power amplifier, respectively. While the power consumption model in (1.12) captures most of the components in Table 1.3, it does not exhibit any dependence on the call traffic load. Experimental results in [48] have pointed out the dependence of the femto-cell BS power consumption on the offered load and the data packet size. Consequently, the power consumption model for a femto-cell BS is expressed by Riggio and Leith [48]

$$P_s = P_d(q, l) + P_f, \tag{1.13}$$

where $P_d(q, l)$ represents the BS power consumption, which depends on the traffic load q [Mbps] and packet size l [bytes], and P_f stands for the idle power consumption component.

In order to capture the temporal fluctuations in the call traffic load, as discussed in Section 1.2.1, a weighted sum of power consumptions at different traffic load conditions (full load, half load, and idle conditions) is considered [7]

$$P_{s,\text{total}} = 0.35 P_{\text{max}} + 0.4 P_{50} + 0.25 P_{\text{sleep}}, \tag{1.14}$$

where P_{max}, P_{50}, and P_{sleep} denote the full rate, half rate, and sleep mode power consumption, respectively. The weights in (1.14) are determined statistically based on the historical traffic records.

Recently, cooperative networking among different BSs and APs in the heterogeneous wireless medium is regarded as an effective approach to enhance the network's overall capacity and reduce the associated energy consumption [1, 4–6, 54]. However, this approach relies on information exchange among different BSs and APs, such as CSI, call traffic load, and resource availability, which are carried mainly over the backhaul connecting these BSs and APs together. Hence, more emphasis is given to the backhaul design and its power consumption. Three types of backhaul solutions can be distinguished, namely copper, microwave, and optical fibre. The most common choice for backhaul is the copper lines [49]. Microwave backhauls are deployed in locations where it is difficult to deploy wired (copper) lines. Also, optical fibre backhauls are mainly used in locations with high traffic due to their high deployment cost. Current research is focusing mainly on the power consumption of microwave and optical fibre links, as they can support the current high data rates. In its simplest form, the microwave (wireless) backhaul power consumption is expressed as [49]

$$P_{\text{BH}} = \frac{C_{\text{req,s}} P_{\text{mw}}}{C_{\text{mw}}}, \tag{1.15}$$

where $C_{\text{req,s}}$ and C_{mw} represent the BS's required backhaul capacity and the microwave backhaul total capacity (100 Mbps), respectively, and P_{mw} denotes the associated power consumption (50 W). However, the model in (1.15) does not account for many features of the backhaul. To gain a better understanding of the power consumption of backhauls, we first provide a brief description of the backhaul structure and associated topologies.

As shown in Figure 1.6, each BS is connected to one or more BSs via a backhaul link. All traffic from BSs is backhauled through a hub node (traffic aggregation point) [55]. Any BS in the network can serve as such a hub node. In general, more than one aggregation level (hub node) can be present. Each hub node is connected to a sink node, which, in turn, is connected to the core network. A BS is equipped with a switch if more than one backhaul link originates or terminates at this BS. Following this description, the microwave backhaul power consumption is expressed as [49]

$$P_{\text{BH}} = P_{\text{sink}} + \sum_{s=1}^{S} P_{\text{BH},s}, \tag{1.16}$$

where P_{sink} is the power consumption at the sink node, $P_{\text{BH},s}$ denotes the power consumption associated with the backhaul operations at BS s, and S stands for the total number of BSs. The following relationships hold

$$P_{\text{BH},s} = P_s(C_{\text{req},s}) + P_{\text{switch},s}(A_s, C_{\text{req},s}), \tag{1.17}$$

$$P_{\text{sink}} = P_{\text{sink}}(C_{\text{req,sink}}) + P_{\text{switch,sink}}(A_{\text{sink}}, C_{\text{req,sink}}), \tag{1.18}$$

where $C_{\text{req},s}$ and $C_{\text{req,sink}}$ represent the required backhaul capacity for BS s and the sink node, respectively. The variable A denotes the number of microwave antennas, P_s and P_{sink} represent the power consumed for transmitting and receiving backhaul traffic for BS s and the sink node, respectively, and P_{switch} models the BS/sink switch power consumption. On the other hand, for an optical fibre backhaul, the power consumption is expressed as [49]

$$P_{\text{BH}} = \left\lceil \frac{S}{\max N_{\text{DL}}} \right\rceil P_{\text{switch}} + S P_{\text{DL}} + N_{\text{UL}} P_{\text{UL}} + \sum_{s=1}^{S} c_s, \tag{1.19}$$

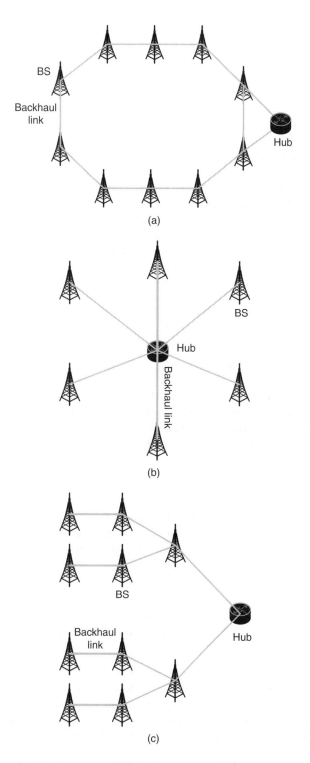

Figure 1.6 Different backhaul topologies [55]: (a) ring topology, (b) star topology, and (c) tree topology

where $\max N_{\mathrm{DL}}$ stands for the maximum number of downlink interfaces available at one aggregation switch, P_{DL} denotes the power consumption due to one interface of a switch, N_{UL} and P_{UL} represent the total number of uplink interfaces and power consumption of one uplink interface, and c_s denotes the power consumption of a pluggable optical interface, which is used to connect a BS to the switch at the hub node.

A limitation with the models (1.9)–(1.16) is that they focus mainly on the BS's operation power. In a more general model, the BS's total consumption is described in terms of the BS's operating energy and embodied energy, E_o and E_e, respectively. The BS's embodied energy represents 30–40% of the BS's total energy consumption [19] and accounts for the energy consumed by all the processes associated with the manufacturing and maintenance of the BS. Over the BS's lifetime, the embodied energy is calculated as 75 GJ [19]. It consists of two components. The first component refers to the initial embodied energy E_{ei}, while the second one stands for the maintenance embodied energy E_{em}. The initial embodied energy comprises the energy used to acquire and process raw materials, manufacture components, and assemble and install all BS components. The initial embodied energy is accounted for only once in the initial BS manufacturing process. The maintenance embodied energy includes the energy associated with maintaining, repairing, and replacing the materials and components of the BS throughout its lifetime. Thus, the BS's total energy consumption (in joules) throughout its lifetime is given by Humar et al. [19]

$$E_b = E_e + E_o = (E_{ei} + E_{em}) + E_o, \tag{1.20}$$

where $E_{em} = P_{em}T_{\mathrm{lifetime}}$, with P_{em} and T_{lifetime} representing the BS's maintenance power and lifetime, respectively. $E_o = P_oT_{\mathrm{lifetime}}$, where P_o is defined in terms of the BS's operating power described by (1.9)–(1.14). The model in (1.20) is useful in quantifying the BS's total power consumption during the network design stage, for example, while designing a multi-tier wireless network. Also, a similar expression can be derived for the backhaul energy consumption in (1.15)–(1.19), which when added to (1.20) can be used to calculate the overall network energy consumption.

1.3.2.2 Mobile Terminal Side

In the literature, different models have been proposed for the MT's power consumption P_m. In the simplest form, P_m captures only the MT's transmission power P_{tm} [29, 46, 50]. To account for the power amplifier efficiency, the MT's power consumption is expressed as [16, 30, 45, 51]

$$P_m = \frac{P_{tm}}{\xi_m}, \tag{1.21}$$

where ξ_m represents the power amplifier efficiency for MT m, $\xi_m \in (0, 1]$. According to this power consumption model, two conclusions can be drawn in terms of the employed modulation and coding schemes (MCS):

- The minimum energy consumption for a data call is attained by using the modulation of the lowest order while satisfying the QoS constraints (e.g. time delay) [28].
- Adopting M-ary frequency shift keying (MFSK) is more energy efficient than adopting M-ary quadrature amplitude modulation (MQAM), since for a given bit error probability, the SNR per bit requirement increases with M for MQAM while it decreases with M for MFSK [20].

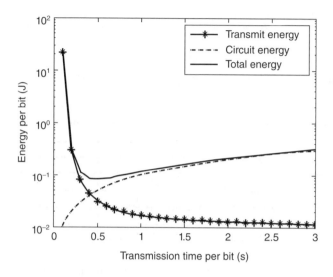

Figure 1.7   MT circuit and transmit energy consumption [56]

However, in practice, the MT circuit's power consumption plays a vital role in the MT's total power consumption, and therefore it should be captured in the power consumption model P_m. Figure 1.7 shows the transmit, circuit, and total energy consumption per bit performance for the MT versus the transmission time per bit [20]. While transmitting using the lowest modulation order, and hence over a long transmission duration, decreases the transmission energy consumption, this is not true for the circuit power consumption. Consequently, the total energy consumption per bit exhibits a minimum value, which corresponds to the optimal MCS. As a result, it is imperative to capture the MT's circuit power consumption in the MT power consumption model. In the literature, three different models have been proposed to represent the MT's circuit power Q_m. The first model assumes that the circuit power consumption is a constant value, independent of the achieved data rate R_m [16, 23, 28, 31, 45, 52]. Such a model changes the aforementioned conclusions regarding the optimal MCS as follows:

- With constant circuit power consumption, adopting the lowest modulation order is no longer the best transmission strategy since energy consumption is directly proportional to the transmission duration [28].
- The optimal MCS is based on the relation between the transmit and circuit power consumption. For long-range applications, the transmission power dominates the circuit power consumption, and hence MFSK is more energy efficient than MQAM, while the opposite is true for short-range applications where the circuit power dominates the total power consumption [20].

One limitation with the constant power consumption model is that it does not reflect the effect of transmission bandwidth and data rate on the MT's circuit power consumption. According to Table 1.4, it is clear that different radio interfaces consume different circuit powers. One reason for such a behaviour is the different operating bandwidths. To account for the effect of the allocated bandwidth, the circuit power consumption presents two terms [53].

Table 1.4 MT power consumption for different technologies [27]

Technology	Action	Power (mw)
WiFi IEEE 802.11 (infrastructure mode)	In connection	868
	In disconnection	135
	Idle	58
	Idle in power save mode	26
	Downloading at 4.5 Mbps	1,450
WiFi IEEE 802.11 (ad hoc mode)	Sending at 700 kbps	1,629
	Receiving	1,375
	Idle	979
2G	Downloading at 44 kbps	500
	Handover to 3G	1,389
3G	Downloading at 1 Mbps	1,400
	Handover to 2G	591

Table 1.5 MT power consumption for different data rates of audio streaming and downloading a 200-MB file using WiFi [27]

Bit Rate (kbps)	Nokia E-71 (mW)	Nexus S (mW)	Samsung Galaxy S3 (mW)
128	990	350	419
192	1,004	390	440
256	1,007	390	452
File download	1,092	998	1,012

The first term represents the digital circuit power consumption, which is modelled as a linear function of the transmission bandwidth (as the bandwidth increases, more computations and baseband processing are required), that is

$$P_{\mathrm{cm}} = P_m^{\mathrm{ref}} + \sigma \frac{B_m}{B_{\mathrm{ref}}}, \qquad (1.22)$$

where P_m^{ref} [W] refers to the reference digital circuit power consumption for a reference bandwidth B_{ref}, and σ denotes a proportionality constant. The second term represents the power consumption of the RF chain and accounts for the power consumption in the digital-to-analog converter, RF filter, local oscillator, and mixer. A limitation of the model in (1.22) is that it does not reflect the effect of transmission data rate on the power consumption, which is evident from Table 1.5. To capture the transmission data rate's impact on the circuit power consumption, a linear function of the achieved data rate is assumed for the circuit power consumption following the fact that the clock frequency of the MT's digital chips scales with the achieved data rate [51]. Consequently, the circuit power consumption is expressed as

$$P_{\mathrm{cm}} = \beta_1 + \beta_2 R_m, \qquad (1.23)$$

where β_1 and β_2 are two appropriately chosen constants, measured in watts and watt per bit per second, respectively. In addition to the transmission and circuit power modelling in P_m, a constant term is introduced to reflect the MT's receiver circuit power consumption [16, 53].

For orthogonal frequency division multiple access (OFDMA) networks, R_m and P_m are defined as the sum of the corresponding terms over multiple sub-carriers assigned to MT m [16, 28, 29, 45, 46, 50]. Similarly, for an MT m enjoying a multi-homing service, R_m and P_m are defined as the sum of corresponding terms over multiple radio interfaces [23, 31].

1.3.3   Energy Efficiency and Consumption Models

Following the aforementioned throughput and power consumption models, we next present several energy efficiency and consumption terms. A summary of these terms proposed in the literature is given in Table 1.6.

A generic definition that can be used regardless of the traffic load condition is referred to as the *energy consumption gain* (ECG), which is defined as the ratio of the energy consumed by a base system (BS, MT, or entire network) to the energy consumed by the system under test, assuming the same conditions [8, 57]. Formally, this is expressed as follows:

$$ \text{ECG} = \frac{E_{\text{base}} - E_{\text{test}}}{E_{\text{base}}}, \tag{1.24} $$

where E_{base} and E_{test} represent the consumed energy measured in joules. The ECG definition in (1.24) is a relative definition that is measured as a percentage. It can be misleading if it is used to compare systems with different characteristics [49].

For access nodes (i.e. BSs and MTs), two definitions can be distinguished. The first definition is referred to as the *energy efficiency index* (EEI), which is defined as the ratio of the attained utility to the consumed energy. On the other hand, the second definition is referred to as the *energy consumption index* (ECI), which represents the reciprocal of the EEI, that is, the ratio of the consumed energy to the attained utility. Overall, both definitions capture the same information; however, they lead to different interpretations [49]. For instance, Figure 1.8 compares the behaviour of an EEI (a) with an ECI (b) [60]. As shown in Figure 1.8a, for the EEI in the low power region, a small improvement in energy saving will lead to a high gain in the EEI; that is, a minor energy saving improvement in a system that is already energy efficient will be interpreted as a high improvement in the achieved energy efficiency. In the medium power region, a high energy saving translates into a small EEI gain, that is, a large energy saving improvement in an energy-inefficient system is interpreted as a small improvement in the achieved energy efficiency. On the other hand, the ECI exhibits a linear relationship between the energy saving improvement and the attained gain in the ECI; that is, a large energy saving improvement for an energy-inefficient system leads to a high gain in the attained ECI. Consequently, for an energy-inefficient system, ECI is more intuitive than EEI. Next, we focus on the EEI and ECI definitions for BSs and MTs subject to different traffic load conditions.

Under low traffic load, it is not required that the BS operates at its full power due to low service demands. Hence, one way to represent the EEI for a given BS at a low call traffic load condition is by means of the ratio between the BS's output power (energy) and the total input power (energy) [2, 22]. That is, the EEI η_s for BS s is expressed as

$$ \eta_s = \frac{P_t}{P_s}, \tag{1.25} $$

where P_t and P_s are the BS's output power (i.e. the power of the RF transmitted signal) and input (consumed) power, respectively. Therefore, η_s is unitless. In addition, at a low traffic

Table 1.6 Summary of different energy efficiency and consumption definitions proposed in the literature [27]

	Model		Comments	References
BS/MT	Energy consumption gain		A ratio of the energy consumed by a base system to the energy consumed by the system under test. It is a relative measure that can be used at any traffic load	[8, 57]
BS	Low traffic load	Output – input power	A ratio of BS output to input power. It is an EEI	[2, 22]
		Area spectral efficiency – input power	The definition measures the power consumed for a certain area coverage. It is used at a low traffic load. It is an EEI	[18, 22]
	High traffic load Network capacity – input power		A ratio of the aggregate BS capacity to the total power consumed by the BS. It is an EEI	[2, 10]
	Temporal fluctuations (ECRW, TEEER, ECRVL)		It uses a weighted sum of power consumption (and throughput, as in ECRVL) at different traffic load conditions. It is an ECI	[7, 49, 58]
	Absolute ECR		It accounts for the absolute temperature. It is an ECI	[7]
MT	Single-user system	Without error consideration	A ratio of throughput to power consumption. It is an EEI	[16, 23, 31, 45, 50]
		With error consideration	A ratio of goodput to power consumption. It is an EEI	[29, 50]
	Multi-user system	Without fairness consideration	It can be the sum rate of all MTs to total power consumption or sum of energy efficiency for individual MTs. It is an EEI	[28–30, 46, 52]
		With fairness consideration	It is the geometric mean of energy efficiencies of all MTs. It is an EEI	[28]
NW	Traffic load independent	APC	A ratio of total power consumption and network coverage area. It is an ECI	[59]
	Traffic load dependent	Rural definition	A ratio of coverage area to power consumption. It is an EEI	[7]
		Urban definition	A ratio of number of users to the total power consumption. It is an EEI	[7]

condition, it is not necessary for the BS to provide a full coverage. It is sufficient to achieve an acceptable coverage probability. As the definition in (1.25) does not capture the BS's achieved coverage, another definition for energy efficiency is proposed to measure the power consumed to cover a certain area [18, 22]. Consequently, the BS's EEI is defined as [18]

$$\eta_s = \frac{T_s}{P_s}. \tag{1.26}$$

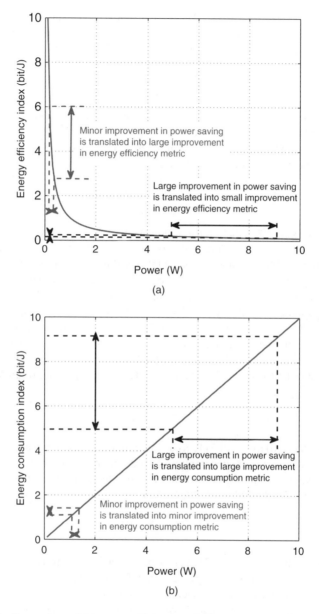

Figure 1.8 Comparison of (a) energy efficiency and (b) energy consumption indices [60]

In (1.26), η_s has the unit of watt^{-1}, and T_s denotes the area spectral efficiency given in (1.6). Under high traffic load conditions, the BS's EEI is defined as the ratio of the aggregate BS capacity to the total power consumed by the BS [2, 10]. Hence, under high traffic load conditions, EEI of BS s is expressed in bits per second per watt as

$$\eta_s = \frac{C_s}{P_s}, \tag{1.27}$$

and C_s is given by (1.5). In (1.25)–(1.27), P_s is usually represented by one definition from (1.9)–(1.13). For MTs, EEI is defined as a measure of the maximum number of bits that can be delivered per joule of consumed energy [16, 23, 31, 45, 50]. EEI is expressed for MT m as

$$\eta_m = \frac{R_m \Delta T}{\Delta E_m} = \frac{R_m}{\Delta E_m / \Delta T} = \frac{R_m}{P_m}, \tag{1.28}$$

where ΔE_m denotes the energy consumed during the time interval ΔT by MT m. However, the expression in (1.28) does not consider the energy consumed for the correct reception of data. Another definition measures the net number of information bits that are successfully transmitted without error per joule [29, 50], and it is expressed as

$$\eta_m = \frac{R_m f(\gamma_m)}{P_m}, \tag{1.29}$$

where $f(\gamma_m)$ represents the packet transmission success rate for a given SNR γ_m for MT m. The expression in (1.29) assumes a ratio of the goodput to power consumption, as compared to the expression in (1.28), which assumes the ratio of throughput to power consumption. The packet transmission success rate $f(\gamma_m)$ follows an S-shaped (sigmoidal) function, exhibiting an increasing trend with respect to γ_m, approaching zero as γ_m approaches zero, and approaching unity as γ_m approaches infinity [29, 50]. The unit of η_m in (1.28) and (1.29) is bits per second per watt. The definitions in (1.28) and (1.29) are proposed for a single-user scenario [16, 23, 31, 45, 50]. In practice, a multi-user system is considered due to the competition over bandwidth [28] and the impact of interference [46] caused by simultaneous transmissions. In a multi-user system, EEI is defined as the ratio between the sum rate of all MTs to the total power consumption [30]

$$\eta_{\text{total}} = \frac{\sum_m R_m}{\sum_m P_m}. \tag{1.30}$$

The definition in (1.30) treats all MTs as a single unit, and takes into account only the total achieved throughput and power consumption. In order to model the system as a set of distinct MTs, an alternative definition is used, which represents the total EEI as the sum of the energy efficiency for each individual MT [28, 29, 46, 52]

$$\eta_{\text{total}} = \sum_m \eta_m. \tag{1.31}$$

The unit of η_{total} in (1.30) and (1.31) is bits per second per watt. However, the definitions in (1.30) and (1.31) do not ensure energy efficiency fairness among different MTs. Therefore, some MTs might exhibit high energy efficiencies while others might present low energy efficiencies very close to zero. To promote fairness among MTs, the geometric mean of energy efficiencies of all MTs is used [28]

$$\eta_{\text{total}} = \sum_m \log(\eta_m). \tag{1.32}$$

Unlike (1.30) and (1.31), η_{total} in (1.32) has a unit of $\log(\text{bps/W})$. In (1.28)–(1.32), R_m is described using (1.7) or (1.8), and P_m is described using (1.21)–(1.23).

While the definitions (1.25)–(1.32) represent the EEI for BSs and MTs at different traffic load conditions, the following definitions are used to represent the ECI. For BSs and MTs, the energy consumption rating (ECR) is defined as the ratio of the power consumption to the achieved capacity [7, 49, 58]. Hence, for BSs, it is the reciprocal of the definition in (1.27), while for MTs, it is the reciprocal of the definitions in (1.28) or (1.29). For BSs, to account for the temporal fluctuation in traffic load, a weighted ECR definition is introduced (ECRW), which assumes the reciprocal of (1.27) while using the weighted sum of power consumption at different traffic load conditions (full load, half load, and idle conditions) to represent P_s as in (1.14) [7, 49, 58]. The telecommunications equipment energy efficiency ratio (TEEER) is calculated as a logarithmic function of ECRW. The ECRW assumes a weighted sum of only power consumption; yet a variable load ECR (ECRVL) definition follows the same expression as ECRW and involves also a weighted sum of the achieved throughput at different traffic loads, using similar weights as in (1.14) [7, 49, 58]. Finally, an absolute ECR definition that accounts for the absolute temperature of the medium, T, is given by Hasan et al. [7]

$$\tilde{\eta}_s = 10 \log \left(\frac{P_s/C_s}{kT \ln (2)} \right), \tag{1.33}$$

where k stands for the Boltzmann constant. According to [7], including the temperature in the analysis follows from the classical thermodynamics theory.

All the aforementioned definitions in (1.25)–(1.33) target access nodes such as BSs and MTs. A definition proposed in the literature for ECI at network level is referred to as the *area power consumption* (APC) [59], which is expressed as the ratio of the total power consumption to the network coverage area, and is measured in W/km^2. The APC is further extended to include the ratio of the total power consumption to achieve a given throughput in a given area, and it is expressed in W/Gbps/km^2. However, for such a metric to be valid in the comparison of different networks, it must be applied to networks with a similar number of sites in a given area [7]. Two EEI definitions can be distinguished at the network level based on the traffic load conditions. The first definition is for rural areas, that is, in a low traffic load condition, and it assumes the expression [7]

$$\eta_n = \frac{\text{Total coverage area}}{\text{Total power consumption at the site}} \ [\text{km}^2/\text{W}]. \tag{1.34}$$

The rationale behind such a definition is that, under low traffic load condition, the main objective is to reduce the total power consumption to cover a specific region. On the other hand, for urban areas with high traffic load, the objective is to reduce the power consumption to achieve a given capacity, and hence the energy efficiency is expressed as [7]

$$\eta_n = \frac{M_{\text{busy hour}}}{\text{Total power consumption at the site}} \ [\text{users}/\text{W}], \tag{1.35}$$

where $M_{\text{busy hour}}$ stands for the number of users in an average busy hour traffic.

1.4 Performance Trade-Offs

Improving energy efficiency of wireless networks is achieved at the cost of some performance degradation. Usually, a threshold level is specified for some target (acceptable) QoS. Green solutions aim to achieve the maximum energy saving while satisfying the QoS threshold. Overall, the performance trade-offs can be divided into two main categories, namely at the network and at the mobile user side, respectively. These trade-offs are discussed next.

1.4.1   Network-side Trade-Offs

The two performance metrics, namely the spectral efficiency and network coverage, conflict with the energy efficiency from a network operator perspective. Both metrics directly affect the network operator's investments, as they are related to the network's available resources in terms of bandwidth (for spectral efficiency) and number and types of deployed BSs (for network coverage).

1.4.1.1   Spectral Efficiency

By definition, spectral efficiency quantifies the system throughput per unit of bandwidth. Such a metric is a key performance indicator for the third-generation partnership project (3GPP) and reflects how efficiently the network bandwidth is utilized in the uplink and downlink. However, the energy efficiency (in the uplink, i.e., from the user perspective, and the downlink, i.e., from the network operator perspective) and spectral efficiency conflict with each other. This conflict is a direct consequence of the relation between bandwidth and power, as illustrated by Shannon's formula in (1.5) for the BSs and in (1.7) for the MTs, respectively. For instance, from Shannon's formula, the relationship between the transmission power and the allocated bandwidth for a given transmission rate is given by

$$P = BN_0 2^{\frac{R}{B}-1}. \tag{1.36}$$

In (1.36), for uplink communications, P stands for the MT's transmission power and B denotes the allocated bandwidth by the BS on the uplink, while for downlink communications, P represents the BS's transmission power and B denotes the allocated bandwidth by the BS on the downlink. The expression in (1.36) is shown in the top left sub-plot of Figure 1.9, which indicates a monotonic relation between the transmission power and the allocated bandwidth. From the top left sub-plot of Figure 1.9, it turns out that, for a given transmission rate R, in order to save transmission power and hence improve the resulting energy efficiency (for the BS on the downlink or for the MT on the uplink), a large transmission bandwidth should be used. In turn, this will reduce the achieved spectral efficiency. Using the energy efficiency definition in (1.27) for BSs or in (1.28) for MTs (and defining P_s or P_m using only the transmission power definitions in (1.9) or (1.21), respectively), the energy efficiency–spectral efficiency relation is expressed as [9]

$$\eta_{EE} = \frac{\eta_{SE}}{(2^{\eta_{SE}} - 1)N_0}. \tag{1.37}$$

The relation in (1.37) is shown in the top right sub-plot of Figure 1.9, and it exhibits a similar performance to the one depicted in the top left sub-plot of Figure 1.9. Hence, η_{EE} converges to the minimum value of $1/(N_0 \ln 2)$ when $\eta_{SE} = 0$, and $\eta_{EE} = 0$ when $\eta_{SE} = \infty$.

However, it should be noted that the expressions in (1.36) and (1.37) account only for the transmission power and do not consider the circuit power component as in (1.10)–(1.14) for the BS and in (1.22) or as in (1.23) for the MT. Accounting for such a circuit power consumption component yields a relationship that exhibits a minimum value, as shown in the bottom left sub-plot of Figure 1.9 for the power–bandwidth relationship and a maximum value in the bottom right sub-plot of Figure 1.9 for the energy efficiency–spectral efficiency relation. From (1.36), it turns out that an infinite bandwidth allocation leads to the minimum

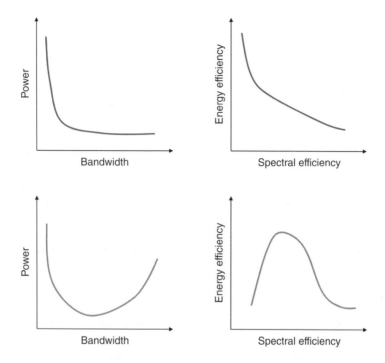

Figure 1.9 Performance trade-offs [9]

power consumption level $N_0 R \ln 2$. However, this is not true for the practical scenario, which involves the device (BS or MT) circuit power consumption. In this case, the green solutions aim to work on the minimum/maximum values in the bottom sub-plots of Figure 1.9 to reach a good power–bandwidth trade-off, and hence a good energy efficiency–spectral efficiency compromise.

1.4.1.2 Network Coverage

Another important metric that conflicts with energy efficiency is the network coverage. High network coverage performance can be achieved in two ways, namely cell stretching and small-cell deployment. The first approach relies on a small number of BSs to cover a large area by stretching the cell coverage as much as possible [9]. While such an approach can reduce the capital expenditure (CAPEX), it leads to high BS transmission power to support MTs at the cell edge. Specifically, it has been shown that for a path loss exponent of 4, the path loss between the BS and cell edge MTs will be reduced by 12 dB if the cell radius is doubled [9]. Consequently, this leads to a 12 dB increase in the BS transmit power to satisfy the target QoS. When only transmission power is considered, energy efficiency scales (degrades) continuously and proportionally with the cell radius. However, accounting for the BS circuit power consumption as in (1.10)–(1.14) leads to a more complex relationship between the cell radius and the achieved energy efficiency. Therefore, green solutions aim to determine the optimal energy efficiency–network coverage (BS cell radius) compromise. The second approach that is adopted to achieve a high network coverage performance relies on small-cell deployment. The main advantage of such an approach is that a low BS transmission power is expected in this case due to the short distance between MTs and the BS. However, it should be noted that using a large number of small-cells might not eventually lead to an improvement in energy efficiency. This is due not only to the BS circuit power consumption but also to the BS embodied energy (1.20). Consequently, green solutions aim to balance the energy efficiency with the network coverage (in terms of specifying the optimal number of small-cells).

Furthermore, the energy efficiency–network coverage trade-off plays a vital role during low traffic load conditions. Specifically, the energy efficiency of wireless networks can be improved by switching off some BSs at a low call traffic load, as will be explained in the next chapter. However, this can result in an increased call blocking probability. Therefore, green solutions aim to achieve energy saving while maintaining the call blocking probability below a certain threshold [12, 26]. In some cases, BSs do not need to be completely switched off, but rather they are allowed to shrink their coverage area by reducing their transmission power, a technique referred to as *cell zooming* [27]. However, this technique may lead to failures in service coverage. Consequently, green solutions aim to enhance the network energy efficiency while maintaining a target performance level in terms of coverage probability $\mathbb{P}_n(\zeta)$.

1.4.2   Mobile User Trade-Offs

The main performance metric conflicting with the energy efficiency from the mobile users' perspective is related to the quality of the ongoing application. Different equivalent measures exist to quantify the quality of the ongoing application, including SNR, data rate, delay (latency), and video quality.

For instance, consider SNR as a QoS indicator [36, 37]. Using the energy efficiency definition in (1.28) for MTs (and defining P_m using only the transmission power definition as in (1.21)), an inverse relationship exists between energy efficiency and SNR, as shown in Figure 1.10a. Specifically, achieving a high SNR requires a high transmission power, which in turn leads to low energy efficiency. However, when the circuit power is accounted for in the MT's total power consumption, as expressed in (1.23), a different relation can be observed, as shown in Figure 1.10b. In this case, two regions can be distinguished in the energy efficiency–SNR relationship. In the first region, the circuit power consumption dominates the MT total power consumption. Consequently, increasing the transmission power (and hence the SNR), will lead to an increased data rate (and hence lower transmission delay). In turn, this will reduce the circuit energy consumption and, as a result, improve the overall energy efficiency. On the other hand, in the second region, the transmission power dominates the MT total power consumption. Consequently, increasing the transmission power (and hence SNR) will lead to high energy consumption and hence reduced energy efficiency (similar to the performance in Figure 1.10a). In practice, the receiver requires a target SNR in order to be able to decode the transmitted signal. Green solutions aim to find the optimal point that balances the achieved energy efficiency–target SNR compromise. A similar argument holds for the BS energy efficiency in the downlink while considering a target SNR threshold.

The same arguments stand for the data rate when it is considered as a QoS metric. When circuit power is dominating the total power consumption, it is more energy efficient to transmit with high power, and hence achieve high data rates. However, as transmission dominates the total power consumption, the high data rate requirement results in reduced energy efficiency. For some applications, a minimum required data rate should be achieved

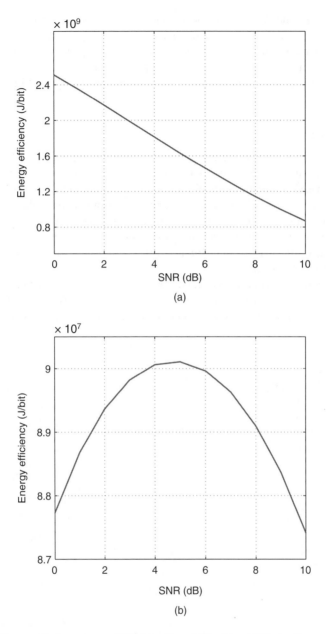

Figure 1.10 Energy efficiency versus SNR (a) with and (b) without MT circuit power consumption

[31, 36, 37] or a constant required data rate should be satisfied [23, 33]. Therefore, green solutions aim to balance the achieved energy efficiency–target data rate compromise.

An equivalent representation to ensure a minimum required data rate is not to violate a maximum delay bound for data transmission. Delay bound has been used as a QoS indicator in [44] for data calls. Similarly, for video streaming applications, in order to maintain a high video quality, video packets should be transmitted before a given delay deadline as in [61, 62]. Stringent delay (latency) requirement calls for high transmission power, which affects the energy efficiency based on which part (circuit power or transmission power) dominates the total power consumption. Green solutions aim to balance the achieved compromise between energy efficiency and target delay bound (video quality).

1.5 Summary

In order to develop and analyse a green network solution, an appropriate definition of energy efficiency and consumption for network operators and mobile users should be adopted. Such a definition is based on the traffic load condition, power consumption, and throughput for network operators and mobile users. In addition, the green network solution should satisfy some target (and possibly conflicting) performance metrics. Therefore, this chapter was dedicated to energy efficiency and consumption definitions, as well as power consumption, throughput, and traffic load models for network operators and mobile users, along with conflicting performance metrics.

After having introduced the necessary background concepts in this chapter, the next chapter will focus on state-of-the-art green network solutions and projects along with the analytical models employed by network operators and mobile users at different traffic load conditions.

2

Green Network Solutions

This chapter mainly presents state-of-the-art green communication solutions and analytical models for both network operators and mobile users at different traffic load conditions. In particular, in green wireless networks, two categories can be distinguished for the proposed solutions and models to enhance and analyse the energy efficiency based on the call traffic load condition. At a low and/or bursty call traffic load, resource on–off switching techniques are adopted, while scheduling techniques are employed at a high and/or continuous call traffic load. In the following sections, green solutions at different traffic load conditions are first reviewed, and then a description of the existing green projects and standards is presented. Future research directions are also presented to address the limitations of the existing approaches.

2.1 Green Solutions and Analytical Models at Low and/or Bursty Call Traffic Loads

On–off switching of radio resources is adopted at low and/or bursty call traffic load conditions to enhance energy efficiency as shown in Table 2.1. Network operators employ on–off switching mechanisms for their BSs at a low call traffic load. Similarly, MTs switch on–off their radio interfaces in a bursty traffic condition. The following sub-sections focus on the related research issues and modelling techniques pertaining to the adoption of these solutions.

2.1.1 Dynamic Planning

Traditionally, the cell size and capacity in network planning are designed based on the peak call traffic load. As discussed in Chapter 1, Section 1.2, the call traffic load exhibits significant spatial and temporal fluctuations. Consequently, the network is over-provisioned at a low call traffic load, which in turn results in energy waste. Switching off some of the available radio resources (e.g. radio transceivers of BSs) at a low call traffic load can yield energy saving and offer acceptable performance. On the contrary, an active BS spends 60% of its total power consumption in processing circuits and air conditioning units (a component represented by the fixed power component in (1.11) and (1.13)) [34]. Consequently, an effective energy-saving approach at a low call traffic load is to switch off some of the network BSs while simultaneously

Green Heterogeneous Wireless Networks, First Edition. Muhammad Ismail, Muhammad Zeeshan Shakir,
Khalid A. Qaraqe and Erchin Serpedin.
© 2016 John Wiley & Sons, Ltd. Published 2016 by John Wiley & Sons, Ltd.

Table 2.1 Summary of green solutions and analytical models at low and/or bursty call traffic loads [27]

Solution/analytical model			Comments	References
BS on–off switching	User association		This phase concentrates the MTs in a few BSs to enable switching off other BSs.	[34, 35, 40, 63–65]
	BS operation		This phase specifies which BSs should be turned off/on and how	[12, 34, 66–70]
MT radio interface on–off switching	With downlink traffic	Without traffic shaping	An MT switches its radio interface if no data packets are available for the MT at the BS	[57, 71, 72]
		With traffic shaping	Traffic shaping at the MT or BS is introduced to enable longer idle duration for the MT	[73–75]
	With uplink traffic		Besides radio interface on–off switching, an MT controls the transmission power, and modulation and coding scheme	[43, 76]
	With bi-directional traffic		This case deals with both uplink and downlink traffic while switching on and off the MT radio interface	[77]

satisfying the target performance metrics. BS on–off switching according to call traffic load conditions is referred to as dynamic planning [12].

In order to design an effective BS switching mechanism, two issues must be addressed, namely the user association problem and BS operation. The BS on–off switching is coupled with the user association problem. In particular, user association is inevitable to concentrate the call traffic load in a few BSs, and hence to switch off other lightly loaded BSs. Therefore, newly incoming MTs should be associated with a subset of active BSs. In addition, MTs already in service should perform handover when their serving BSs are switching off. Two research directions related to MT association can be identified. The objective of the first direction is to develop new energy-efficient user association mechanisms [34, 35, 63–65], while the second direction aims to derive analytical models to assess the performance of different energy-efficient association mechanisms [40].

In developing an energy-efficient MT association mechanism, two approaches can be adopted to meet the MT target QoS while concentrating the call traffic load in a few BSs. The first approach assumes an objective function that minimizes the networks' energy consumption while accounting for the user target QoS constraints. On the contrary, the second approach aims to balance the trade-off between MTs' flow-level performance (e.g. data rate or delay) and network energy consumption [35]. The latter case assumes a multiobjective optimization problem with a weighting factor. When the weighting factor equals zero, the MT association is determined based only on the MT flow-level performance. As the weighting factor increases, the MT association decision pays more attention to the network power consumption performance. When the weighting factor reaches infinity, the MT is associated to the BS that maximizes the network energy efficiency performance in bits per Joule. Overall, the MT association mechanism can assume a centralized or decentralized architecture [34]. The objective of both architectures is to concentrate the traffic load in a few BSs while satisfying the MTs' target data rate and the bandwidth limitations of BSs. The centralized mechanisms use a central controller that uses global network information related to channel conditions and user requirements to perform an energy-efficient MT association. On the contrary, in a decentralized architecture, an MT locally selects the BS with the highest call traffic load that can serve its target data. The main challenge in designing such a decentralized mechanism is the associated computational complexity imposed by the binary nature of the BS on–off switching decision variables, and hence the mixed-integer nature of the optimization problem. Consequently, greedy algorithms are mainly used to reach a good (sub-optimal) switching decision [34, 35]. In order to design greedy algorithms, a decision criterion should be defined. For instance, a greedy algorithm based on a user–BS distance decision criterion switches off the BSs with the longest user–BS distance to improve the network energy efficiency [63]. The rationale behind this decision criterion is that the longer the user–BS distance, the greater the transmission power required to satisfy the users' target service quality. Another decision criterion is referred to as the network impact, which quantifies the impact of switching off a given BS on the network performance [63]. Switching off a given BS leads to additional load increments into the neighbouring BSs. In addition, switching off a BS can also lead to a positive effect on the neighbouring BSs because of the reduced inter-cell interference. By quantifying the two aforementioned measures, the network impact criterion maps the switching off decision as a BS selection problem, whose objective is to find the BSs that when switched off result in the highest network impact [64]. Furthermore, the coverage holes represent an important problem associated with BS on–off switching. As a result, another BS on–off switching decision metric is related to coverage holes avoidance. In [65], it has been shown that finding the optimal set of BSs that: (i) minimizes the network power consumption and (ii) avoids coverage holes is closely related to the minimum-weight disc cover problem. This problem is known to be NP-hard, and a greedy algorithm is proposed to switch off BSs while maintaining network coverage in polynomial time complexity.

Queueing models are proposed in the literature to assess the performance of different energy-efficient MT association mechanisms [40]. The energy-efficient MT association process in the overlapped coverage of different BSs is modelled as a customer joining a queue with $|\mathcal{S}| \times |\mathcal{M}|$ servers, where $|\mathcal{S}|$ and $|\mathcal{M}|$ denote the number of BSs with overlapped coverage and the maximum number of MTs that can be accommodated in each BS, respectively. For instance, consider a two-BS scenario with three service areas. In service areas 1 and 2, an MT is served by the BS covering that area. In service area 3, an MT can be served by either BS with overlapped coverage. A BS is switched off, and consequently, its corresponding $|\mathcal{M}|$ servers are shut down, if no MT is assigned to it. Following such a queueing model, analytical expressions are derived for call-blocking probability, average number of MTs assigned to each BS and average power and energy consumed by the network operator to serve one MT [40]. The model can be further approximated to account for the multiple-BS overlapped coverage case.

On the basis of the MT association phase, the BS operation decision is specified. In particular, BSs with a concentrated call traffic load become active, while lightly loaded BSs are switched off. The BS operation phase deals with three concerns: (i) accommodating future traffic demands, (ii) determining BS wake-up instants for switched-off BSs and (iii) implementing the BS on–off switching decisions. For accommodating future traffic demands, it should be noted that the BS operation decision lasts for a long duration (i.e. several hours), since frequent BS on and off switching is not desirable due to the increased energy consumption in the BS start-up phase [12] and service unavailability for the off cells during the decision computation phase [34]. Consequently, the BS operation decision should account for the future call traffic load by reserving some resources (bandwidth) to accommodate the future demands [34] by exploiting the past call traffic load patterns to estimate the future load [12]. Another approach to estimate the future traffic demands is based on an online stochastic game [66], where neighbouring BSs communicate with each other to predict their traffic profiles, leading eventually to optimal switching decisions and minimum network energy consumption.

For determining the BS wake-up instants, it should be noted that switching off some cells is acceptable only if the active BSs extend their coverage areas to support the cells with inactive BSs. When the call traffic load of the inactive cells increases beyond the capacity limit of the active BSs, some of the inactive BSs should be switched on. Therefore, besides specifying which BSs to be switched off, another equally important research issue refers to determining the wake-up instants for switched-off BSs. Two BS wake-up schemes are proposed in [67], namely the number M- and the vacation time V-based schemes, respectively, as shown in Figure 2.1. In the M-based scheme, the BS is switched off in an idle condition (i.e. no MT in service) and it wakes up when M users arrive at the BS coverage area. On the contrary, for the V-based scheme, the BS remains in a sleep state for a specific period of vacation time before waking up. Two versions can be distinguished for the V-based scheme, namely the single vacation and multiple vacations. In the single vacation case, the BS remains awake after the vacation period even if there is no call request to serve, while in the multiple vacations case, the BS goes back to sleep if it wakes up and finds no call request to serve. A limitation associated with the M-based scheme is that the BS needs to continuously monitor the user request arrivals, which translates into an advantage for the V-based scheme. For femto-cell BSs in overlapped coverage with macro-cell BSs, three wake-up modes can be distinguished in the literature, namely BS-controlled, MT-controlled and network-controlled modes [68]. In the BS-controlled mode, the femto-cell BS performs continuous sensing for user activity to wake up, while in the MT-controlled mode, the MT sends wake-up messages for a sleeping femto BS. In the network-controlled mode, the core network controls the femto-BS operation through wake-up messages over the backhaul link. The three different wake-up modes result in different performance in terms of BS and MT energy consumption and signalling overhead. The BS-controlled mode leads to less energy saving for the BS. The MT-controlled mode increases the energy consumption for the MT, while the network-controlled mode incurs additional signalling overhead [68]. Markov decision process (MDP)-based optimal wake-up schemes are presented in [69] for network-operated femto BSs overlapping with a macro BS. In order to wake up the right femto BSs, which serve the extra traffic load and still lead to efficient energy usage, call traffic load and user localization within the macro-cell information are required. In the absence of the traffic localization information, the femto-BS wake-up problem is formulated as a partially observable MDP [69].

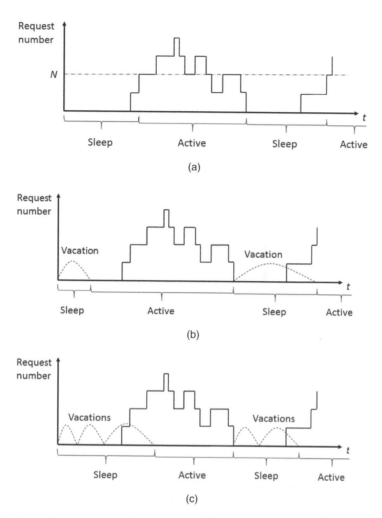

Figure 2.1 BS wake-up schemes (a) M-based scheme; (b) V-based scheme: single vacation; (c) V-based scheme: multiple vacations [27]

The last issue dealt with in the BS operation problem addresses the switching off mode entrance and exit stages, which are two important design stages in implementing the BS operation decision [70]. The switching off mode entrance stage specifies how the transition from the on (active) state to the off (inactive) state is implemented. If a BS is switched off very fast, the corresponding MTs may not be able to successfully execute their handover procedures and their calls eventually will be dropped. This could be due to a strong received signal from the BS that serves the MT, which prevents the MT from hearing signals from nearby BSs. Hence, if the BS that an MT is associated with is suddenly switched off, the latter will not be able to synchronize and connect to another active BS. Another reason is the maximum number of handovers that can occur simultaneously towards a new BS, due to the limited signalling channel capacity. Hence, a progressive switching off operation, that is, referred to as BS wilting [70],

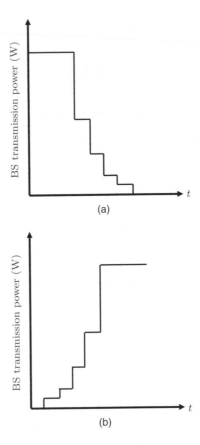

Figure 2.2   BS switching off mode entrance and exit [27]. (a) BS wilting; (b) BS blossoming

can be used, as shown in Figure 2.2. In BS wilting, the BS transmission power is progressively
halved until the BS is switched off. During this process, the MTs associated with the wilting
BS initiate a handover process to the neighbouring BSs and the BS switching off procedure is
suspended in case of unsuccessful handover of MTs. On the contrary, the switching off mode
exit specifies how the transition from the off (inactive) state to the on (active) state is implemented. A BS that is switched on too fast can generate a strong interference to MTs in service.
As a result, a progressive switching on process, that is referred to as BS blossoming [70], can
be used as shown in Figure 2.2. In BS blossoming, the BS transmission power is progressively
doubled until the BS is switched on.

2.1.2   MT Radio Interface Sleep Scheduling

Similar to BS on–off switching (dynamic planning), an MT with a bursty or low traffic load
can save energy by switching off its radio interface from time to time. An appropriate on–off
switching (sleep) schedule design for the MT radio interface varies based on whether the MT
establishes communications on the downlink [57, 71–75], uplink [43] and [76] or both links
[77].

For downlink communications, two categories can be identified for the MT radio interface on–off switching mechanism based on whether a traffic-shaping technique is employed or not [57, 71–75]. In the absence of traffic–shaping techniques, the MT radio interface switching off decision (in case of a bursty or low traffic load) is based on the unavailability of data packets for the MT at the serving BS. Consequently, the MT switching on–off (sleep) schedule specifies the switching off intervals and switching on instants for the radio interface based on the data availability. At a switching on instant, the MT checks if there are any packets available for it at the serving BS. In the absence of data packets, the MT enters a switching off interval; otherwise, the MT keeps its radio interface active to receive the available packets. During the MT sleep interval, all incoming data packets are buffered at the BS until the next MT switch on instant. On the one hand, a long sleep interval can enhance the MT energy savings; however, it also increases the packet-buffering delay at the BS until it is received by the MT. Furthermore, buffer overflow at the BS will result in discarding future incoming data packets for the sleep MT. In addition, unnecessarily switching on the MT radio interface to check for data packet availability at the BS buffer leads to MT energy losses. Thus, the main research objective in this case is to design a sleep schedule for the MT radio interface that maximizes its achieved energy saving while reducing the buffering delay of data packets available at the BS. One approach in designing such a schedule is by modelling the MT radio interface as a server that assumes repeated vacations [57, 71], as shown in Figure 2.3. Following this queueing model, analytical expressions can be derived for the expected number of sleep intervals until a data packet is available for the MT at the BS. Using these analytical expressions, myopic optimization problems can be formulated to minimize the energy consumption rate of the MT while achieving an acceptable message response time performance, where the message response time is defined as the time interval from the arrival time of an arbitrary message (data packet) at the BS to the time the message (data packet) leaves the system (BS) after service completion [57]. In addition to myopic optimization techniques, dynamic programming can be used to design a sleep schedule that minimizes a cost function consisting of a weighted sum of the MT energy consumption with radio interface on–off switching and a target performance metric (e.g. the buffering delay at the BS for the MT when its radio interface is switched off) [71]. Besides queueing models coupled with myopic and dynamic optimization techniques, a Llyod-max algorithm can be used to design a sleep schedule that specifies the switching on instants for the MT radio interface [72].

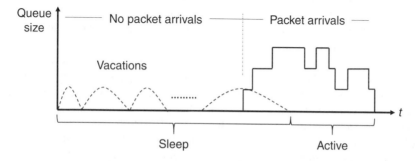

Figure 2.3 Modelling of MT on–off switching as a server with repeated vacations [27]. The model is similar to the BS V-based scheme with multiple vacations

The main limitation with the works in [57, 71] and [72] is that if the packet inter-arrival duration of the application is too small, the MT will not be able to switch off its radio interface to provide an acceptable QoS performance. Moreover, the MT consumes a significant amount of energy to switch on its radio interface. Every time the MT finds a single data packet available at the BS buffer, an interruption signal is triggered by the MT radio interface to activate the data bus and central processing unit (CPU) of the MT. Consequently, the MT will not be able to enter into a deep sleep state if it experiences frequent interrupts, and hence only a small amount of energy will be saved. Therefore, traffic-shaping techniques are employed to enable a longer idle duration for an MT. In this context, the idle duration denotes the interval during which an MT is not receiving any data packets. The traffic-shaping technique can be implemented by the MT by buffering the incoming data packets at its radio interface for a short period, without activating the data bus and CPU of the MT. Then, the data packets are released as a burst, which reduces the interruption-triggering events, and therefore, more energy is saved [73]. For transmission control protocol (TCP)-based applications, an alternative approach can be adopted by the MT, where the BS is forced to send data packets in bursts, and hence, it enjoys a longer idle duration, by announcing a zero congestion window size. Thus, the data packets are buffered for a longer period at the BS, until an appropriate window size is announced by the MT to allow the BS to release the data packets in bursts [74]. While the aforementioned traffic-shaping research deals with a single-user environment, the main goal in a multiuser scenario is to schedule the sleep intervals of the radio interfaces for different MTs to satisfy their target QoS and achieve energy saving by switching off the radio interfaces of MTs for a long enough duration [75]. An MT stores in its buffer sufficient data to satisfy its QoS and then switches off its radio interface for energy saving while the BS serves another MT. The MT activates its radio interface only when the data available at its buffer are insufficient to satisfy the required QoS.

For uplink communications, in addition to adapting the physical layer parameters as in controlling the MT transmission power and modulation/coding schemes, an MT can further save energy by switching on and off its radio interface. In [76], it is argued that different parameters such as the packet arrival rate and packet delay constraint have a significant impact on the practicality of employing such a switching approach. An on–off switching mechanism can be employed for energy saving at MTs for applications with small packet arrival rates and/or large packet delay constraints. In these scenarios, the research objective is to jointly adapt the power control, modulation and coding schemes (MCS) and switching on and off strategies of the MT radio interface to save energy in accordance with the stochastic traffic and channel conditions (i.e. no a-priori knowledge of traffic arrivals and channel conditions). In such a case, an MT is able to switch off its radio interface and hold data packets in its buffer to transmit them in bursts in better channel conditions. In addition to saving energy, the transmission mechanism should avoid an overflow event at the MT buffer and satisfy the required QoS in terms of data packet delay. An MDP problem can control the data packet transmission throughput (and hence, the amount of buffered data packets at the MT), resulting bit error probability, and MT radio interface state (switch on or off) to balance energy saving with QoS guarantee (i.e. minimizing data packet delay and avoiding buffer overflow) [43].

A general model for MT radio interface sleep scheduling is presented in the context of bidirectional communications in [77]. In this scenario, incoming downlink traffic does not suffer from BS-buffering delays during uplink transmissions, since the MT radio interface is already active. Thus, a finite general Markov background process can be used to model both the uplink activity and downlink traffic to derive analytical expressions for the buffer occupancy and downlink packet delay statistics [77]. These expressions are useful in developing an efficient on–off switching mechanism for the MT radio interface for both uplink and downlink communications.

2.1.3   Discussion

Following the above review, BS on–off switching (dynamic planning) aims to exploit spatial and temporal fluctuations in the call traffic load to achieve energy saving. Consequently, adopting static call traffic models in the switching schedule design (i.e. to determine the switch off and wake-up instant decisions) and/or performance evaluation, as in [65], is not realistic. The call traffic load models should capture the joint spatial and long-term temporal fluctuation behaviours [12, 63]. On the contrary, traffic models [34, 35] that reflect joint spatial and short-term temporal call-level fluctuations are incapable of assessing the daily switching schedule performance due to time-varying traffic demands. In addition, the traffic models that reflect only the long-term (as in [63] and [66]) or short-term (as in [40]) temporal call-level fluctuations fail to exploit the spatial dimension of the problem, and are unrealistic for performance evaluation studies in large-scale networks with multiple BS sites. For BS power consumption models, both static and dynamic components, as in (1.11) and (1.13), should be accounted for, which is the case for the algorithms developed in [12, 35] and [65]. On the contrary, the power consumption models that assume constant transmission power, as in [34, 40, 64] and [66], neglect the transmission power scaling associated with the call traffic load and represent unrealistic models. Overall, the reported solutions in Section 2.1.1 aim to minimize the network energy consumption, which is somehow similar in concept to maximizing the energy consumption gain given in (1.24). However, this expression does not assess the network gain (in terms of transmitted power as in (1.25) or network coverage as in (1.26)) versus the incurred cost (in terms of the network-consumed power). The reported solutions minimize the network energy consumption while satisfying a target performance metric. For BS on–off switching solutions, the trade-offs are based on admission quality requirements (i.e. network coverage and call blocking) as in [12, 34, 40] and [65]. Few works account for mobile user trade-offs [35]. From a practical perspective, a solution should account for both network and mobile user trade-offs to better serve the users who required QoS [63, 64]. Green solutions that adopt BS on–off switching report 25–50% energy savings [14, 78].

Similarly, MTs can save energy by switching off their radio interfaces during idle periods of bursty traffic. Thus, static traffic models that assume a fixed number of backlogged data packets ready for transmission [76] are not realistic to determine the MT idle periods, and therefore, will not be useful in developing practical sleep schedules for the MTs. The practical traffic models should reflect the packet-level short-term temporal fluctuations [43, 57, 71–75] and [77]. While some solutions account for both active and idle power consumption values [57, 71, 72] and [74] and reception power consumption [75] and [77], such solutions do not account for the MT circuit power consumption component. Both transmission and circuit power consumptions should be accounted for [43] and [76]. Yet, these models assume fixed MT circuit power consumption and neglect the dynamic circuit power component as described in (1.22) and (1.23). The reported solutions in Section 2.1.2 minimize the MT energy consumption while accounting for the mobile user trade-offs. However, such a modelling approach overlooks the network capacity limitations, for example, in terms of available bandwidth, which may lead to call blocking. Therefore, the proposed solutions should account for both network and user trade-offs.

2.2   Green Solutions and Analytical Models at High and/or Continuous Call Traffic Loads

Energy-efficient scheduling techniques are adopted at high and/or continuous call traffic loads to satisfy the target QoS with reduced energy consumption when the on–off switching techniques are infeasible. In the literature, various scheduling techniques have been proposed for network operators and mobile users. These scheduling techniques can be divided into five categories, as shown in Table 2.2. These categories include scheduling for single-network access, multi-homing access, small size cells, relaying and Device-to-Device (D2D) communications and scheduling with different energy supplies. These topics will be next addressed.

2.2.1   Scheduling for Single-Network Access

In this technique, a mobile user receives the required resources from a single wireless access network at a time. In the literature, two system models are adopted for single-network access. The first model deals with a single network that covers a given geographical region, and it is referred to as a homogeneous wireless medium. The second model assumes the availability of multiple networks with overlapped coverage in the geographical region, and it is referred to as a heterogeneous wireless medium.

In the homogeneous wireless medium, the network operator assigns radio resources to MTs in a way that reduces the total power consumption of its BSs. This objective can be achieved by minimizing the BS transmission power while providing acceptable QoS performance for the MTs, a technique that is referred to as the margin-adaptive strategy [10]. One approach to implement the margin-adaptive strategy adopts a score-based scheduler. For instance, in an OFDMA system, the BS calculates a score for every radio resource block q to be assigned to MT m [79]. Such a score q ensures that the BS consumes the least transmission power by allocating the resource block q to MT m. In addition, the score promotes a fair resource assignment among MTs, since a penalty function is included based on the number of already assigned resource blocks for MT m. A low-score q reflects a more desirable resource block. Fairness issues are also studied in [80] following a proportional rate constraint that ensures each user eventually receives a specific proportion of the system throughput. Admission control policies are also used to implement a margin-adaptive strategy, where a new session (call) is admitted into the system as long as the sub-frame energy in an OFDMA-based BS is kept below a certain threshold [81]. Furthermore, a margin-adaptive strategy can be implemented following a discrete rate adaptation policy that controls both the transmission rate and power according to the channel conditions to maximize the achieved energy efficiency for a target bit error rate [82]. Similarly, a channel-driven rate and power adaptation strategy can be implemented by jointly adapting the MCS and transmission power to optimize the trade-off between goodput and energy efficiency [83]. Moreover, a margin-adaptive strategy can be implemented via resource scheduling among MTs based on their traffic delay tolerance [38]. Delay-tolerant traffic (e.g. video and data) can be opportunistically served during periods of good channel conditions (i.e. soft real-time service). A drawback associated with the margin-adaptive strategy

Table 2.2 Summary of green solutions and analytical models at high and/or continuous call
traffic loads [27]

Solution/analytical model				Comments	References
Single-network	BS	Margin adaptive strategy		It minimizes the transmission power while providing an acceptable QoS	[10, 38, 79–83]
		User association in heterogeneous medium		It assigns the MTs to the BSs to save energy	[32, 84–87]
	MT	OFDMA network	Sub-carrier allocation	It exploits joint sub-carrier allocation and power control	[29, 36, 46]
			Carrier aggregation	It employs both PCC and SCC for energy saving	[42]
		TDMA network		It exploits opportunistic transmission	[44]
Multi-homing	BS	Network cooperation		The MT receives the required data rate from multiple BSs simultaneously. The BSs coordinate their transmitted power for energy saving	[33, 36]
	MT	BS selection and power allocation		The MT specifies a set of BSs for uplink transmission and determines the allocated transmission power for each radio interface	[23, 31]
Relays and D2D communications		Small-cells		It divides the cell into several tiers of smaller cells to reduce the transmission range for BSs and MTs	[41, 88, 89]
		Relays		Fixed and/or mobile relays are used to reduce transmission distance	[7, 14, 90–96]
		D2D		Mobile nodes in close proximity communicate with each other directly	[97–104]
Multiple energy sources	BS	Multiple retailers		The network operator decides how much electricity to procure from each retailer	[2, 105]
		On-grid and green energy sources		The objective is to maximize the utilization of green energy and to save the on-grid energy	[37]
		Complementary renewable sources		The BSs are powered using only renewable sources	[106–114]
	MT	Multiple batteries		It employs the recovery effect of batteries	[95]

is that it requires CSI knowledge to allocate the transmitted power, which further necessitates the use of pilot symbols. Such pilot symbols will incur some energy consumption. In the literature, two approaches can be used for pilot energy assignment [10], namely the constant single pilot energy and the constant total pilot energy. In the former approach, each pilot maintains the same energy level independent of the number of pilot symbols. Thus, the larger the number of pilot symbols, the more accurate the CSI, and yet higher energy is consumed. The later approach allocates fixed energy to all pilots, which leads to reduced energy per pilot for a larger number of pilots, and to inaccurate CSI.

On the contrary, in a heterogeneous wireless medium, energy can be saved by assigning MTs to the BSs that reduce energy consumption for the set of network operators with BSs that assume overlapped coverage [32]. Moreover, in such a heterogeneous environment, each BS may choose between two modes of operation, namely point-to-point and point-to-multipoint mode. Thus, the problem can be decomposed into two sub-problems, namely the BS selection and BS operation mode selection. While the work in [32] controls the transmission power only through the BS operation mode selection, a joint BS selection and power control mechanism is proposed in [84] to associate MTs to BSs with overlapped coverage with the aim to minimize the BS transmission power to reduce the interference among different communication channels. In addition, data offloading techniques can be adopted to improve energy efficiency in a heterogeneous wireless medium. In particular, through mobility prediction and using the pre-fetching feature, data traffic can be offloaded from cellular networks to WiFi hotspots and femto cells [85]. Consequently, delay-tolerant traffic can be downloaded when mobile users are close to the WiFi access point or femto cell rather than using the macro cell [86]. Overall, data offloading can be either network- or user-driven [87]. Various factors, such as user mobility, backhaul throughput, data size and WiFi and/or femto-cell densities affect the energy efficiency performance when data offloading is adopted [85].

Similarly, MTs can save energy by appropriate uplink radio resource scheduling, based on the network multiple access scheme. In the literature, various energy-efficient mechanisms are proposed for OFDMA-based networks [28, 29, 42] and [46]. The proposed mechanisms mainly improve energy efficiency through sub-carrier allocation, power control and joint sub-carrier allocation and power control [29]. Centralized and decentralized architectures can be adopted to implement the radio resource allocation mechanisms [28, 46]. In a centralized architecture, the BS in each cell jointly performs sub-carrier allocation, modulation order adaptation and power control for the MTs in the uplink. In a distributed mechanism, given a sub-carrier assignment, an MT adjusts its modulation order and transmission power to optimize its own energy efficiency. In a multicell environment, multicell interference should be taken into account while designing an energy- efficient uplink resource allocation scheduling [29, 46]. In addition to sub-carrier allocation and power control, energy efficiency is maximized for OFDMA-based networks through dynamic carrier aggregation [42]. Although an MT served by all carrier components will enjoy an enhanced throughput, its energy consumption also increases. In a dynamic carrier aggregation technique, an MT is assigned to the queue of a given carrier component that is referred to as the primary carrier component (PCC). Whenever the queue of a given carrier component is empty, it helps other carrier components through aggregation, and therefore, it is referred to as the supplementary carrier component (SCC). Two mechanisms can be employed for SCC assignment [42]. The first mechanism aggregates all SCCs to support the PCC with the longest queue. The second mechanism orders PCCs according to the queue length, and SCCs are circularly allocated to the ordered PCCs in a round-robin manner.

For time division multiple access (TDMA)-based networks, energy efficiency is maximized for a set of MTs via opportunistic transmission [44]. A scheduler is considered at the BS to select an MT for transmission and determine its transmission rate. The problem complexity is reduced by decomposing it into two tasks. The first task is a user scheduling sub-problem that opportunistically selects an MT for transmission, based on the channel conditions and backlog information. The second task specifies the transmission rate for the selected MT to minimize the transmission power by transmitting data packets in queue such that the average delay constraint is satisfied with equality.

2.2.2 Scheduling for Multi-Homing Access

Currently, the wireless communication medium is a heterogeneous environment with overlapped coverage due to different networks. In this networking environment, MTs are equipped with multiple radio interfaces. Using the multi-homing capability, an MT can maintain multiple simultaneous associations with different wireless access networks. In addition to enhancing the achieved data rate by bandwidth aggregation, the multi-homing service can improve the energy efficiency of network operators and mobile users as MTs experience different channel conditions and bandwidth capabilities over their different radio interfaces.

By supporting multi-homing services, different network operators can reduce the transmission power of their BSs. The reason behind this reduction can be explained using the concept of power–rate curve, which can be graphically divided into two regions [33]. In the first region, the transmission power increases slowly with the growth of the data rate, while in the second region the transmission power increases dramatically with the data rate. Thus, a multi-homing data rate threshold, R_b, can be specified to enable multi-homing transmission if the required data rate is higher than R_b [33]. The data rate multi-homing threshold relies on the ratio of the channel gain between the MT and the BSs of different networks. Moreover, the optimal transmission data rate from each BS can be determined to maximize the networks' energy efficiency. In addition, cooperating BSs can control their transmission power by following a semi-Markov decision process (SMDP) to minimize the total power consumption of the BSs under a target QoS constraint at the MTs [36].

Similarly, MTs can improve their energy efficiency via the multi-homing service. In this case, an MT specifies how many and which BSs will be selected for multi-homing, according to the required data rate and the channel parameters of available BSs [23]. In order to reduce the associated complexity, the resource allocation problem is decomposed into two sub-problems. The first sub-problem determines which BSs will be selected for multi-homing, while the second sub-problem specifies the optimal transmission rate for each selected BS. For a constant data rate service, the energy efficiency maximization task is equivalent to the MT total power consumption minimization problem. Different from [23], the authors in [31] deal with energy efficiency maximization for a variable data rate using power allocation in a multi-homing service.

2.2.3 Scheduling with Small-Cells

Small-cell (e.g. pico and femto cells) have a radio coverage in the range of tens to a few hundred metres [88]. Consequently, the division of a macro-cell into several tiers of smaller cells replaces a long transmission range with a short transmission range because of the close proximity between small-cell BSs and MTs [88]. The small-cell power consumption is expected to reach approximately 5 W by 2020 [89]. Thus, an improved energy efficiency can be achieved by small-cell deployment. In [88], an expression is provided for the possible power gain $G(J)$ resulting from the macro-cell splitting into J smaller cells. It is shown that for an ideal free space propagation channel model, the achieved power gain satisfies $G(J) < 1$, which means that cell splitting should not be implemented in this case. On the contrary, in a non-ideal propagation environment, $G(J) > 1$ and it increases with the number J of small-cells, that is, the power gain improves with the number of deployed small-cells. However, it should be noted that the BS power consumption model in [88] does not capture the BS-embodied energy as in (1.20). When the BS-embodied energy is accounted for, there is a limit on the number of small-cells that can be deployed to improve energy efficiency. In the literature, different configurations have been adopted for the small-cell deployment, as shown in Figure 2.4. The cell-on-edge deployment mainly distributes the small-cells around the edge of a macro cell to serve the cell-edge users. On the contrary, in the uniform deployment, the small-cells are uniformly distributed across the macro cell. In [89], it is shown that the cell-on-edge deployment leads to a significant reduction in the network energy consumption as compared with the uniformly distributed configuration, due to the lower transmission power for cell edge users.

The main challenge for adopting a cell-splitting approach is the associated inter-tier interference. This is due to the limited available radio resources. As a result, the macro-BS radio resources are shared among the small-cells. Multicell processing can be employed to mitigate the resulting interference [88]. Thus, multiple BSs within a cluster exchange CSI and users' data to serve MTs and eliminate the associated interference. Using the gathered information, beam-forming techniques are applied to minimize the total transmission power while ensuring a certain signal-to-interference plus noise ratio (SINR) for different MTs. Besides multicell processing (and in the presence of both co-tier and cross-tier interference), admission control with QoS guarantee plays a vital role in mitigating interference, where a joint radio resource allocation mechanism can be employed among the multitier networks [41].

2.2.4 Relaying and Device-to-Device Communications

Another approach to reduce the transmission distance and hence achieve energy saving for BSs in the downlink and MTs in the uplink is relaying. In this regard, two types of relays can be employed [7, 14]. The first type is based on fixed relay stations, which are defined as network elements (repeaters) that store and forward the data towards the destination. The second type utilizes MTs as relays and it eliminates the cost of installing fixed relays in the network, yet it increases the system complexity.

Fixed relays are very useful in cell coverage extension and in reducing the power consumption of BSs (in the downlink) and MTs (in the uplink) due to a short transmission range. However, in this regard, the fixed relay stations should not be confused with the small-cell deployment. A small-cell mainly acts as an independent BS that decodes the mobile users' information and passes the decoded information together with the signalling information to the network operator via wired backhaul links [90], as shown in Figure 2.5. On the contrary, the relays mainly forward the user information from MT to BS and vice versa. Two types of fixed relays can be distinguished according to the 3GPP LTE-advanced and IEEE 802.16j standards, as shown in Figure 2.5. A type I relay can help an MT, which is located out of the

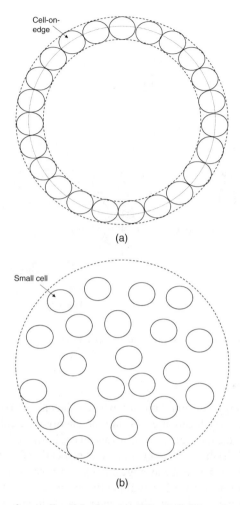

Figure 2.4   Configurations for small-cell deployment [27]. (a) Cell-on-edge deployment; (b) uniformly distributed deployment

coverage area of a given BS, to access that BS, while a type II relay can help an MT within the coverage area of a given BS to improve its service quality with reduced power consumption.

Overall, two research directions can be identified in the context of relaying for green networking when fixed relays are employed. The first research direction deals with the optimal placement of fixed relay stations to promote energy efficiency in wireless networks. The optimal relay placement for green networking is mainly affected by several key parameters including the distance between the relay station and the nodes (BSs and MTs), the radio propagation environment and line-of-sight conditions, the relay height and the relay coding scheme (e.g. amplify-and-forward and full/partial decode-and-forward). The authors in [91] propose a geometrical model for energy-efficient relay placement while accounting for the aforementioned factors. In addition, these authors identify the maximum cell coverage of a relay-assisted cell

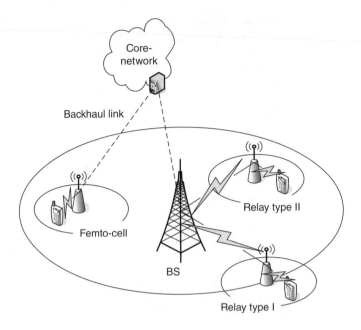

Figure 2.5   Illustration of the difference between the relay station and femto-cell

and the average cell energy consumption. It has been shown that (i) for a given relay coding scheme, an optimal relay location exists and the energy efficiency performance rapidly degrades away from this location and a more advanced coding scheme is required to maintain a good performance and (ii) there exists an optimal relay location for which increasing the cell coverage has a minimal impact on the average energy consumed per unit area. In addition to optimizing the relay position, the work in [92] jointly optimizes the relay position and its serving range for energy-efficient operation. The second research direction mainly deals with relay assignment and radio resource allocation for energy-efficient operation. It has been shown that the maximum benefit of the system can be achieved if the single best relay is selected for a specific source–destination pair [93]. Hence, for a set of source–destination pairs, the objective is to assign the optimal relays (and their respective radio resources, e.g. power and sub-carriers) to the source–destination pairs to enhance the network energy efficiency [94].

Installing fixed relays incur additional infrastructure, operational and maintenance costs for the network operators. In addition, each single relay consumes power for data forwarding and hence high power consumption is expected in a wireless network with dense relay deployment [95]. Consequently, employing MTs as mobile relays is a more appealing solution for network operators. In this context, MTs with good channel conditions can forward the data between BSs and other MTs in both directions (uplink and downlink). The main research challenge is how to motivate rationale (selfish) MTs to act as relays while relaying in turn will cost them additional power consumption. One direct approach to stimulate MTs to act as mobile relays is through a payment system. Hence, a buyer–seller market scenario might be adopted between the source nodes (MTs in uplink or BSs in downlink) and the relaying MTs. In order to reach an agreement on the amount of utilities a buyer pays and the set of radio resources the seller offers, a double auction game theory can be used [96].

In addition to using MTs as relay nodes, D2D communications can be employed to reduce the transmission range and hence achieve energy saving in wireless networks. While the MT only forwards the data between source (BS or MT) and destination (MT or BS) nodes in mobile relaying, MTs in close proximity directly communicate with each other. In this context, three types of D2D communications can be distinguished, namely in-band underlay, in-band overlay and out-band communications, respectively. In in-band underlay, both the cellular and D2D communications use the same resources. Hence, the main objective is to allocate resources among cellular users and D2D users in a way such that D2D users do not create interference to the cellular users. In [97–100] and [101], power efficiency is maximized through joint power allocation and mode selection for the MTs. Hence, MTs can choose between cellular and D2D communications, and the mode that maximizes power efficiency for the wireless network is selected. Such a mode selection results in a binary decision, which leads to a mixed-integer program formulation due to the real-valued nature of the power allocation problem and the binary nature of the mode selection problem. Consequently, heuristic algorithms are proposed to reduce the associated computational complexity [100, 102]. In in-band overlay D2D communications, dedicated resources are assigned to cellular and D2D users. In [103], the authors proposed a BS-assisted D2D in-band overlay communication algorithm that can enhance the network energy efficiency. In particular, in the peer discovery phase, instead of relying on the MT beacon signals (which incur power consumption), BSs assist the D2D users to identify their peers. Furthermore, radio resources (e.g. power and bandwidth) can be allocated to cellular and D2D users in a more energy-efficient manner and via a less complex approach. In out-band D2D communications, the D2D communications take place over a separate band than the cellular band, for example, WiFi band. For instance, in [104], the authors proposed to form clusters for the cellular users who are in close proximity for WiFi communications. MTs coordinate their D2D communications over the cellular radio interface while exchanging their data over the WiFi direct interface. High energy efficiency is achieved as MTs can select one of the two modes: cellular or D2D communications.

2.2.5 Scheduling with Multiple Energy Sources

Various scheduling techniques have been proposed in the literature to deal with the presence of multiple energy sources [26, 37, 105–114] and [115]. The objective of these works is to simultaneously control transmission power and select the energy source that minimizes the total energy consumption. For network operators, multiple energy sources address the availability of different electricity retailers [26, 105], on-grid and green (renewable) energy [37] and different (complementary) renewable sources [106–113] and [114]. For MTs, multiple energy sources consider the availability of multiple batteries at the MT [115].

In an electricity market liberalization model, electricity retailers compete with each other to achieve the highest individual profits by adjusting the electricity price offered to users in different regions [26]. The electricity prices offered by different retailers change frequently to reflect the variations in the cost of energy supply, a strategy which is referred to as real time pricing. For a set of electricity retailers, a Stackelberg game can be formulated, where each retailer provides the real time price to maximize the own profit, and the network operator determines how much electricity to procure from each retailer to power on its BSs and achieve the lowest call blocking with the least monetary cost [26]. In [105], the optimal amount of energy to be procured from each retailer is specified via evolutionary algorithms (i.e. genetic algorithm and particle swarm optimization), which are shown to outperform the deterministic algorithm [26] because of the random nature of the evolution process. In addition to the presence of multiple electricity retailers, it is argued that the BSs of future cellular networks will be powered by both on-grid and green (renewable) energy (e.g. solar and wind energy) [37]. Hence, hybrid energy systems are expected to power the future BSs, where a combination of renewable and grid energy sources is utilized, as shown in Figure 2.6. Complementary renewable energy sources can be employed as well. If the power grid is absent, that is, the BS is not connected to the power grid (and hence, the controller 2 in Figure 2.6 does not exist), the BS is powered only by the renewable sources. An energy-harvesting battery should be used, as shown in Figure 2.6, to overcome the intermittent nature of renewable energy sources. With such a hybrid energy system, the objective is to optimize energy utilization in such networks by maximizing the green energy utilization and saving of on-grid energy. In this case, network designers are faced with the following two main concerns [37]: (i) how to optimize the usage of green energy at different time slots to accommodate the temporal dynamics of the green (solar) energy generation and the call traffic load and (ii) how to accommodate the spatial dynamics of the call traffic load with the objective of maximizing the green energy utilization by balancing the green energy consumption among BSs through cell size adjustment. While the aforementioned works assume the presence of on-grid energy, the long-term objective is to power BSs in appropriate locations using only a combination of complementary renewable sources (e.g. wind in winter and solar in summer) [106]. Moreover, cooperative techniques enable different BSs (networks) to share (trade) their green power, whenever possible, with each other for a sustainable and energy-efficient network operation [107].

In order to use renewable energy sources, renewable energy generation and storage should be investigated. As renewable energy sources are intermittent, energy storage units are

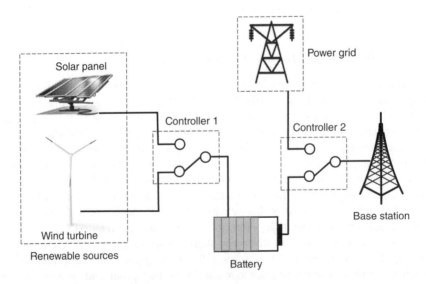

Figure 2.6   Green hybrid solution [27]

deployed to address this limitation. Thus, the harvested (solar, wind) energy is stored in a battery with finite capacity before it is used in data transmission [108, 109]. In this context, the energy replenishment process and the storage constraints of the rechargeable batteries should be taken into account while designing energy-efficient transmission strategies [110]. Two constraints should be considered at the energy-harvesting battery [111]. The first constraint ensures that the energy drawn from the battery is almost equal to the energy stored in the battery, a condition which is referred to as the causality constraint. The second constraint ensures that the energy level at the battery does not exceed a maximum level to avoid battery energy overflow. Consequently, storage sizing is very important to guarantee a sustainable energy at a reduced monetary cost. Moreover, BSs have to adapt their data transmission to the energy available at a particular time instant [112, 113]. Therefore, more studies are needed to minimize the overall power consumption of BSs through on–off switching at a low call traffic load or through scheduling and node cooperation [114] at a high call traffic load to reduce the required energy generation and battery storage capacity. A very important aspect of green communications is to consider the environmental dimension of the proposed solution. For selecting an appropriate energy supply (i.e. electricity retailer and/or renewable energy source), it is necessary to guarantee that the associated CO_2 emission cost is below a target level. The CO_2 emission cost, in kg/h, related to the BS power consumption P_s is given by Bu et al. [26]

$$I(P_s) = \alpha P_s^2 + \beta P_s, \tag{2.1}$$

where α and β are constants that depend on the pollution level of the electricity retailer.

For MTs that adopt a pulsed discharge profile, the battery is able to recover some charges during the interruptions of the drained current (i.e. no transmission period). Thus, an improved battery performance can be achieved. This phenomenon is referred to as the recovery effect. In order to promote the recovery effect and enhance the battery performance (and hence improve energy efficiency), a package of multiple batteries can be used and a scheduling policy can be developed to efficiently distribute the discharge demand among the multiple batteries connected in parallel [115].

2.2.6 Discussion

The majority of research works that investigate green communication solutions at a high traffic load assume static traffic models for radio resource scheduling and performance evaluation [10, 26, 28, 29, 32, 33, 46, 79, 80, 84, 88] and [105]. Very few works in the literature employ traffic models that reflect long term (as in [81] and [89]) or short term (as in [38, 41] and [42] for call-level and [44] for packet-level) temporal fluctuations. Also, few works assume traffic models that capture spatial fluctuations in traffic load [37] and [36]. Spatial and temporal traffic models should be employed for performance evaluation of green resource-scheduling algorithms. Spatial traffic models are useful in evaluating the algorithm performance in large-scale networks, while temporal models are important to investigate the associated signalling overhead, which may jeopardize the energy-saving benefits, if high overhead is expected. Moreover, many references account only for transmission power consumption [10, 29, 32, 33, 38, 41, 42, 81, 84, 88] and [89]. Both transmission and circuit power consumption should be considered [26, 28, 36, 37, 44, 46, 79, 80] and [105]. However, the aforementioned models do not account for dynamic circuit power consumption, as depicted by (1.22) and (1.23). In addition,

BS transmission power consumption should scale with the traffic load as expressed in (1.11) and (1.13). Also, for small-cell and multi-tier deployment, both the operation and embodied energy should be accounted for as in (1.20) to avoid misleading conclusions. While some works aim to minimize energy consumption, reference [79] aims to maximize an energy consumption gain expression similar to (1.24). Furthermore, the works in [28, 29, 33, 42, 46] and [80] aim to maximize an energy efficiency expression similar to (1.27), (1.28) or (1.29). Such an expression provides a better indication of the performance in terms of the achieved gain (the resulting data rate) versus the incurred cost (the energy consumed). Almost all reported solutions aim to minimize the energy consumption or maximize energy efficiency, while maintaining a satisfactory performance that balances the mobile user operation. The works in [26] and [105] aim to balance the network operator objectives. In practice, an effective solution should account for both network operator and mobile user [41].

Green solutions that adopt small-cell deployment reports up to 60% energy savings when combined with BS on–off switching [7, 14] and [78]. Other green solutions that employ renewable energy to power BSs report up to 40% power savings [78]. In what concerns the green solutions that exploit relay deployments, they report 5–20% energy savings [14]. Finally, green solutions based on D2D communications report an improvement of 20–100% in terms of power efficiency [116].

2.3   Green Projects and Standards

Due to the environmental and financial consequences of high energy consumption in the telecommunications industry, several projects were launched in the United States, Europe and Japan to investigate energy-efficient technologies in wireless networks. Sample projects include GreenTouch [117], EARTH [118], OPERA-NET [119], Mobile VCE [8] and Green IT [120], which will be briefly described next.

The GreenTouch consortium was launched in the period January 2010–January 2015 and was led by Alcatel-Lucent/Bell Labs with many collaborators (in total 30 operators and manufacturers) from academia and industry. The main objective of the GreenTouch consortium is to provide energy-efficient techniques to reduce the energy consumption of the Information and Communication Technology (ICT) sector by 1,000 folds. Towards such an objective, the consortium covers all the network components including the core network (switching and routing) and the wireless and mobile front ends (BSs and MTs). The GreenTouch consortium investigated research problems related to sustainable data networks, optical networks and large-scale antenna systems.

The EARTH (Energy Aware Radio and neTwork tecHnologies) project was launched by the European Commission in the period January 2010–June 2012 and was funded with 15 million Euros [58]. The project was led by European mobile operators and research organizations with the objective of reducing the energy consumption of mobile networks at least by 50%. The project mainly covered four research directions. The first direction mainly addressed energy efficiency metrics at the system level. The second direction targeted energy-efficient architectures such as cell size optimization, heterogeneous network deployment and adoption of relay and cooperative communications strategies. The third direction dealt with energy-efficient radio resource management such as dynamic load management and transmission mode adaptation, joint power and resource (bandwidth or time slot) allocation, interference management and multiradio access technology coordination. The last direction investigated radio access technologies including multiple-input-multiple-output (MIMO), adaptive (reconfigurable) antennas and power control at component, front end and system level.

The OPERA-NET (Optimizing Power Efficiency in mobile RAdio NETworks) project was led by France Telecom in the period June 2008–May 2011 in response to the European Union's concerns towards the environmental impacts of the ICT high energy consumption. The project was funded by 5 million Euros and targeted four research directions. The first direction defined the key performance indicators (KPIs) for energy efficiency and investigated energy saving in BSs via dynamic planning. The second direction studied optimization techniques for link-level energy efficiency and energy-aware device (BSs and MTs) design. The third direction focused on technology enablers such as developing power amplifiers with high efficiency. The last direction dealt with the mobile radio access network's end-to-end energy efficiency.

The mobile VCE (Virtual Center of Excellence) was a long-term project launched in United Kingdom in two phases and was funded by industry and government. The first phase started in 1997, while the second phase took place in the period January 2009–2011. The objective of the mobile VCE project was to reduce energy consumption in high-speed networks by hundred folds. The project mainly focused on reducing energy consumption in BSs and MTs at component level (via power amplifiers and processors), dynamic planning and sleep modes, relaying and radio resource management. One contribution of the project was the introduction of the class J power amplifier, which offers efficiency in the range of 85–90% [14]. The Green IT project was launched in Japan and involved 100 companies and research institutes. The project mainly targeted power efficiency in data centres and networks and set regulations and mechanisms to encourage green networking.

In addition to the aforementioned projects, much effort of standardization took place towards promoting green networking. For instance, the 3GPP promoted new modifications to the BS management mechanisms [78]. In particular, such modifications introduce the basic signalling procedure to switch on or off a given BS through its backhaul interface. For UMTS, this has been established using the I_{ub} interface between the BS and the radio network controller (RNC) and using the X2 interface in the LTE networks. Furthermore, 3GPP is currently investigating standardization of the signalling information exchange among different networks in the heterogeneous wireless medium for cooperative energy-saving solutions as discussed in Sections 2.2.1 and 2.2.2.

Despite the existing solutions and the ongoing research and standardization effort, many issues still remain unanswered for developing green wireless networks. These open research topics are discussed in the next section.

2.4 Road Ahead

The existing research mainly focuses on improving the energy efficiency of either network operators or mobile users. However, a green solution implemented at the network operator side can result in high energy consumption at the mobile user side and vice versa. Therefore, green solutions should capture the trade-off in energy efficiency among network operators and mobile users and should be jointly designed to balance such a trade-off.

For instance, the BS on–off switching mechanism involves two phases, namely user association and BS operation. Targeting only energy efficiency of the network operator, a BS on–off switching mechanism can result in an energy-inefficient user association from the mobile user perspective. In particular, it can result in MTs being associated with a far-away BS in the uplink to switch off a nearby BS. This will lead to energy depletion for the MTs, and thus to dropped services. Therefore, a BS on–off switching mechanism should capture the trade-off in the achieved energy efficiency for the network operator and mobile users, and should aim at balancing them. MTs should be associated with BSs that can balance energy saving for both network operators and mobile users. However, the existing research targets balancing energy consumption performance of a BS with the flow-level performance at the MT [35]. The multiobjective function in [35] should aim to balance the energy saving for both BSs and MTs while satisfying the MT required QoS. Consequently, the BS switching off decision criteria such as User–BS distance [63], call traffic load [34], network impact [64] and network coverage holes [65] should be revised. The switching off criterion should include, besides the aforementioned metrics, an MT energy consumption metric. Similarly, the existing mechanisms employ only the call traffic load as a wake- up criterion [69]. The switching mechanisms should capture the degradation in energy consumption for MTs and should include it as a BS wake-up decision metric. Moreover, MTs suffer from inter-cell interference. An uplink-scheduling scheme at MTs performs power allocation while handling the inter-cell interference negative effect. However, inter-cell interference can be affected by the BS on–off switching decision. Such a dependence can be modelled in the user-received SINR using a BS activity parameter, which is equal to one if the BS is on, and zero otherwise. In turn, the BS on–off switching decision should promote energy saving at MTs by switching off cells that lead to the highest interference during a low call traffic load condition. Furthermore, the analytical models used in the literature, for example, the queueing model in [40], mainly assess the network energy-saving performance for a given mechanism. These models should be extended to assess the energy-saving performance for both network operators and mobile users.

Similarly, the existing MT radio interface on–off switching mechanisms focus mainly on the energy-saving performance at the MT without capturing the impact of the implemented energy-saving mechanisms. In particular, the downlink mechanisms enable an MT to switch off its radio interface for a given interval while dealing with only the buffer delay and/or overflow at the BS, for example, [57, 71–74] and [75]. However, the impact of BS on–off switching is not considered while taking the switching off decision at the MT. If the serving BS is switched off during the MT sleep interval, the MT connection will be dropped and the buffered data will be lost. Consequently, the MT radio interface sleep-scheduling algorithm needs to be revised. For instance, in [57], the MT switching on is triggered upon a packet arrival at the BS. This model should be extended to account for the BS switching off decision as an additional switching on trigger for the MT radio interface. Furthermore, the existing switching off design metrics focuses on balancing energy consumption at the MT with the buffer delay at the BS [71]. An extension is required to account for the BS energy consumption due to a delayed switching off decision for the BS while waiting for the MT to become active. Moreover, network operators can save energy at BSs by scheduling delay-tolerant applications (e.g. data and video) opportunistically in the presence of good channel conditions. MT radio interface sleep scheduling should take account of the delay at the BS due to both MT inactivity and BS opportunistic scheduling of traffic. The radio interface on–off scheduling at an MT and the opportunistic traffic scheduling at the BS should balance energy efficiency at both network operators and mobile users, while satisfying the target performance metrics. For the MT energy-saving mechanisms at the uplink, power control and radio interface on–off switching mechanisms account in their design only for the channel and traffic dynamics [43]. Besides the aforementioned dynamics, the BS on–off switching dynamics should be captured while designing an energy-saving mechanism.

In addition, the energy-efficient radio resource-scheduling mechanisms at a heterogeneous wireless medium assign MTs to the BSs that reduce energy consumption for network operators [32] and [84]. These mechanisms mainly target downlink resource scheduling. However, no investigation is performed for MTs with bidirectional traffic, for example, video call applications. In this case, two approaches can be implemented to achieve energy saving at both network operators and mobile users. The first approach relies on single-network access, where the MT is associated with the BS that balances energy saving for the network operators and the mobile users. On the contrary, the second approach employs multi-homing, where the MT connects on the uplink to the BS that promotes energy saving for the mobile user while the MT connects on the downlink to the BS that promotes energy saving for the network operators. Furthermore, the potentials of the heterogeneous wireless medium should be better exploited to enhance energy saving. For multi-homing service, as MTs connect to multiple networks simultaneously, the radio resources at different radio interfaces can be properly scheduled to enhance energy efficiency. The existing research works focus only on power allocation schemes at the different radio interfaces of MTs to save energy in various channel conditions. Given the bandwidth capabilities of different networks, cross-layer designs that incorporate joint bandwidth and power allocation can lead to enhanced energy efficiency.

Furthermore, the existing opportunistic scheduling mechanisms focus on energy saving for network operators [38] or MTs [44]. However, for MTs with bidirectional traffic, opportunistic scheduling should be implemented such that the time slot for uplink and downlink transmission can balance energy savings for both network operators and mobile users. Finally, for radio resource scheduling in BSs powered by renewable energy sources, the existing research focuses mainly on downlink delay-tolerant applications [37] and [105]. Thus, BSs aim to schedule data transmissions at time slots when energy is available. However, when MT radio interface on–off scheduling is implemented, the BSs need to account for the MT sleep interval, which may conflict with the BS energy limitation due to the finite size of the energy-harvesting buffer at the BS and might result in buffer overflow. Consequently, the resource-scheduling mechanism should balance energy availability at the BS with energy saving at the MT.

Finally, the existing D2D communication approach does not fully exploit the presence of multiple radio interfaces at the MT and the multi-homing capability. In the literature, an MT can establish a direct link for D2D communications only over the cellular radio interface for in-band communications [121, 122] and [123]. In addition, an MT can use the cellular radio interface for coordination while using another radio interface (e.g. WiFi direct or Bluetooth) for data transmission for out-band communications [124] and [125]. In both cases, data transmission takes place only over a single link between a D2D pair. Enabling data transmissions over multiple radio interfaces in D2D communications can take advantage of the diverse resources available at different radio interfaces (e.g. the supporting bandwidth). Aggregating such radio resources at the sink device allows for an improved system performance in terms of the achieved throughput, latency and energy efficiency, and it represents a strategy that requires further investigation.

2.5 Summary

Given the call traffic load condition, different green solutions and analytical models can be adopted. At a low call traffic load condition, on–off switching of radio devices (e.g. BSs for network operators and MT radio interfaces for mobile users) can improve the performance of energy consumption. Radio resource-scheduling techniques have been proposed for a high call traffic load condition. Despite the various effort proposed to analyse and design effective green solutions, many open issues remain to be further investigated. As future research, green solutions should capture the trade-off in energy efficiency among network operators and mobile users, and should be designed to balance such a trade-off.

On the basis of the background information provided in Chapters 1 and 2, the next chapters investigate some of the open research issues listed in Section 2.4. In particular, the second part of the book, Chapters 3–6, addresses green multi-homing radio resource allocation solutions. Chapter 3 introduces the green multi-homing radio resource allocation fundamentals and limitations of existing research. Chapters 4–6 investigate the green multi-homing radio resource allocation problem for the downlink radio frequency (RF) heterogeneous medium, uplink RF heterogeneous medium and downlink RF and visible light communication (VLC) inter-networking, respectively. The third part of the book, Chapters 7–10, is dedicated to network management solutions such as dynamic planning (Chapter 7), cell-on-edge deployment (Chapter 8) and D2D communications (Chapters 9 and 10).

Part Two

Multi-homing Resource Allocation

3

Green Multi-homing Approach

The wireless communication medium has become a heterogeneous environment with overlapped coverages due to the coexistence of different cells (macro, micro, pico and femto), networks (cellular networks, WLANs and WMANs) and technologies (RF, VLC and D2D communications). In such a networking environment, MTs are equipped with multiple radio interfaces. Through multi-homing capability, an MT can maintain multiple simultaneous associations with different networks. Besides improving the achieved data rate through bandwidth aggregation, the heterogeneous wireless medium together with the multi-homing service can boost energy efficiency of network operators and mobile users. This is mainly due to the vast diversity in fading channels and propagation losses among MTs and BSs, available resources and operating frequency bands of different networks. This chapter mainly focuses on promoting energy efficiency in wireless networks through multi-homing resource allocation, exploiting network cooperation and integrating different and new network technologies such as RF and VLC. In such a network setting, a joint bandwidth and power allocation approach, for uplink and downlink communications, can promote energy savings for both mobile users and network operators. In this chapter, we first present the heterogeneous wireless communication medium, then discuss the potentials of multi-homing approach in green communications and focus on several challenging issues pertaining to the design and implementation of the green multi-homing approach.

3.1   Heterogeneous Wireless Medium

At present, the wireless communication medium is a heterogeneous environment with overlapped coverage from different networks (cellular networks, WLANs and WMANs), cells (macro, micro, pico and femto) and technologies (RF, VLC and D2D communications) [1]. Despite the fierce competition in the wireless service market, the aforementioned wireless networks, cells and technologies will continue to coexist due to their complementary service capabilities. For instance, while the IEEE 802.11 WLANs can support high data rate services in hot spot areas, the cellular networks and the IEEE 802.16 WMANs can provide broadband

Green Heterogeneous Wireless Networks, First Edition. Muhammad Ismail, Muhammad Zeeshan Shakir,
Khalid A. Qaraqe and Erchin Serpedin.
© 2016 John Wiley & Sons, Ltd. Published 2016 by John Wiley & Sons, Ltd.

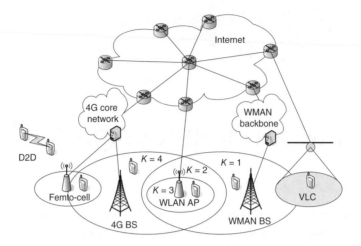

Figure 3.1   Illustration of a heterogeneous wireless network [126]

wireless access over long distances and serve as a backbone for hot spots [12]. The basic components of such a heterogeneous wireless communication network are the wireless networks, MTs and a core Internet protocol (IP)-based network [3], as illustrated in Figure 3.1. The wireless networks allocate radio resources to MTs to equip them with connectivity via the IP-based core network. Next, we describe the heterogeneous wireless network components, including the wireless networks and MTs with emphasis on the available radio resources and propagation attenuation [126].

3.1.1   Wireless Networks

The heterogeneous wireless medium is composed of a set $\mathcal{N} = \{1, 2, \ldots, N\}$ of wireless networks, which can have different access technologies, such as the fourth-generation (4G) cellular networks, WLANs, VLC and WMANs, as illustrated in Figure 3.1. While 4G cellular networks, WLANs and WMANs rely on different RF access technologies, VLC relies on visible light for communications. Particularly, in VLC networks, communications take place by modulating the intensity of light in light-emitting diodes (LEDs) in such a way that it is not observed by the human eyes. Consequently, the illumination energy that is consumed for lighting is also used for communications. More details about VLC networks, the challenging issues in integrating VLC with RF networks and radio resource allocation in an integrated VLC and RF network for green communications are provided in Chapter 6. Chapters 3 through 5 mainly focus on the integration of RF networks for green communications. The cellular networks in $\mathcal{N}$ can have different cell sizes, for example macro, micro, pico and femto cells. In such a networking environment, different networks are assumed to be operated by different service providers. Also, different networks operate in different radio frequency bands. Every network has a set of BSs or access points (APs), $\mathcal{S}_n = \{1, 2, \ldots, S_n\}$ for network n, to provide service coverage, radio resource management, user mobility management and access to the Internet.

With overlapped coverage from the BSs/APs of different networks, different service areas can be distinguished. The set of service areas is denoted as $\mathcal{K} = \{1, 2, \ldots, K\}$. Each service area, $k \in \mathcal{K}$, is covered by a unique subset of BSs/APs of various networks. For instance, as shown in Figure 3.1, service area 2 is covered by the cellular network and the WMAN, while service area 3 is covered by all the three networks.

Network n ($\in \mathcal{N}$) BS/AP s ($\in \mathcal{S}_n$) has a maximum available bandwidth of $B_{ns,\max}$. A cooperative networking environment is considered, where the BSs/APs of different networks are connected through a backbone that enables them to exchange their signalling information.

3.1.2 Mobile Terminals

The networking environment assumes a set $\mathcal{M} = \{1, 2, \ldots, M\}$ of MTs. Network n BS/AP s has a subset of MTs $\mathcal{M}_{ns} \subset \mathcal{M}$ that lie in its coverage area. Two modes of communications can be distinguished for each MT. The first mode is based on the communication between MT m with network n BS/AP s. The second mode relies on direct communications among MTs in close proximity based on D2D communications. This part of the book (Chapters 3 through 6) mainly focuses on green communications among MTs and BSs/APs. More details about D2D communications and their application to green communications are provided in Chapters 9 and 10.

Each MT is equipped with multiple radio interfaces, and assumes a multi-homing capability for multiple simultaneous associations with different networks. Hence, a given application of the MT, for example, data downloading/uploading or video streaming, can be served on the downlink/uplink using multiple networks. Such an infrastructure presents the following advantages [4]. First, using the multi-homing capability, the available resources at different networks can be aggregated to support bandwidth-hungry applications through multiple threads at the application layer. Second, multi-homing calls enable better mobility supports to ensure that at least one of the used radio interfaces will remain active during the call. Third, the multi-homing support can reduce the call-blocking rate and improve the system capacity.

Each MT $m \in \mathcal{M}$ has its own home network, but it can also get service from other available networks at its location, using the multi-homing capability. The subset of MTs $\mathcal{M}_{ns}$ in the coverage area of network n BS/AP s is divided into the subset $\mathcal{M}_{ns1}$ of network subscribers whose home network is network n and the subset $\mathcal{M}_{ns2}$ of network users whose home network is not network n.

3.1.3 Radio Resources and Propagation Attenuation

Let B_{nsm}^{UL} (B_{nsm}^{DL}) denote the allocated bandwidth from network n BS/AP s to MT m on the uplink (downlink), where the allocated bandwidth is equal to 0 for $m \notin \mathcal{M}_{ns}$. Let P_{nsm}^{UL} (P_{nsm}^{DL}) denote the allocated transmission power for communication between MT m and network n BS/AP s on the uplink (downlink). In addition, let P_{nsm}^{c} denote the circuit power that is required to keep the radio interface active for communication between MT m and network n BS/AP s, with $P_{nsm}^{\mathrm{UL}} = P_{nsm}^{\mathrm{DL}} = P_{nsm}^{\mathrm{c}} = 0$ for $m \notin \mathcal{M}_{ns}$. The maximum power is denoted by P_{mT} for MT m, and by $P_{ns,\max}$ for network n BS/AP s.

The channel power gain between MT m and network n BS/AP s on the uplink (downlink) is denoted by h_{nsm}^{UL} (h_{nsm}^{DL}), and it captures both the channel fast fading and path loss. Let the distance between MT m and network n BS/AP s be d_{nsm} and the path loss exponent be

α. Hence, the path loss between MT m and network n BS/AP s is given by $d_{nsm}^{-\alpha}$, where $d_{nsm} > d_f$ and d_f denotes the far-field distance of the transmitting antenna. The one-sided noise power spectral density is denoted by N_0.

3.2   Green Multi-homing Resource Allocation

In the literature, various energy-efficient (green) radio resource allocation schemes have been proposed to satisfy the mobile user-required QoS at reduced energy consumption. Two categories of solutions can be distinguished in the heterogeneous networking environment, for single-network access and multi-homing access, respectively, as discussed in Chapter 2. The single-network access radio resource allocation schemes do not fully exploit the available resources in the heterogeneous wireless medium, where an MT can communicate through multiple radio interfaces with different channel conditions and radio bandwidths to enhance the energy efficiency. In multi-homing access, an MT connects to all available BSs/APs of different networks, and radio resources are allocated to enhance energy efficiency in the networking environment.

In the literature, multi-homing energy-efficient (green) radio resource allocation mechanisms, for example, [33], mainly deal with how to optimally allocate uplink/downlink transmission power by adapting to different channel conditions and path losses among MTs and BSs/APs, given an allocated bandwidth. In addition to exploiting channel conditions and path losses from different BSs/APs in a heterogeneous wireless medium, other resources can be used to enhance energy efficiency, such as the available bandwidths and different operating frequency bands at different BSs/APs. Hence, the resulting energy efficiency can be further improved through a joint bandwidth and power allocation approach [126]. Different MTs can be allocated different amounts of bandwidths on the uplink (downlink), based on the MT (BS) maximum power and channel conditions, which will affect the associated transmission power allocation.

While the joint bandwidth and power allocation approaches have been investigated in the literature for OFDMA networks in terms of joint sub-carrier allocation and power control, the existing works are mainly limited to single-network access. As a result, the solutions proposed for OFDMA networks cannot be directly applied to the heterogeneous networking environment due to the following facts. First, in an OFDMA single-network access, the networking environment does not present different service areas with service coverage from unique BSs/APs. Second, in a heterogeneous wireless medium, the MT is located at different distances from the BSs/APs of different networks (and hence suffers from different path losses), which affects the resource allocation decision differently from the single-network access, where an MT is served by only one BS/AP. Third, in a single-network access, the resource allocation decision does not need coordination among BSs/APs of different networks, which is different from the heterogeneous networking environment. Hence, the heterogeneous networking settings have to be taken into consideration in developing a joint bandwidth and power allocation approach to maximize energy efficiency.

Consider an uplink communication scenario in a heterogeneous networking environment, as illustrated in Figure 3.2 via MTs 1 and 2. The MTs are located at different distances from the three BSs and experience different channel conditions. In addition, the three BSs have different available bandwidths. In this scenario, the MT with low battery energy and/or bad channel conditions (e.g. MT 1) can be allocated larger bandwidth by the BSs (BSs 1 and 2) than

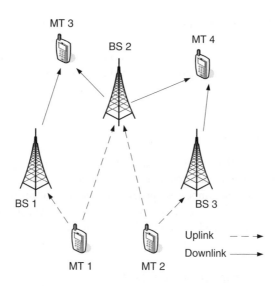

Figure 3.2 Illustration of multi-homing uplink and downlink radio communications in a heterogeneous wireless medium [126]

the MT with high available battery energy and/or better channel conditions (e.g. MT 2 which obtains its required bandwidth from BSs 2 and 3) to satisfy the required QoS of both MTs, via a multi-homing service. This in turn can reduce the energy consumption for the MT suffering from low available energy and/or bad channel conditions (MT 1), which leads to improved energy efficiency for the networking environment. The objective of resource allocation is to enhance the energy efficiency of the MTs, while satisfying their required QoS. In this context, the main factors affecting the resource allocation decision are the MT available energy and required QoS, and the available bandwidths at the BSs, together with the channel conditions.

Similarly, for a downlink communications scenario, as illustrated in Figure 3.2 using MTs 3 and 4, the BSs/APs can allocate their available bandwidths to MTs to increase the energy efficiency of the different networks. Unlike the uplink communications scenario, the main factors affecting the resource allocation decision include the BS/AP maximum transmission power in addition to their available bandwidths and MTs' required QoS, together with channel conditions.

The joint bandwidth and power allocation approach can be regarded as a cross-layer design since bandwidth is allocated at the network layer from the BSs/APs of different networks, while the transmission power (and circuit power for MTs) is allocated at the physical layer. Such a cross-layer design can implement opportunistic communication over the wireless links for a set of MTs with different channel conditions and available energy to enhance energy efficiency of the networking environment. This is possible through backward and forward information flows between the network and physical layers. Such a cross-layer design spans not only different layers but also different entities, that is, BSs/APs of different networks and MTs. Coordination is required among different layers and entities (MTs and networks on one side and different networks on the other side) in the joint bandwidth and power allocation approach. Different networks should coordinate their allocated resources (bandwidth in uplink and downlink and transmission power in the downlink) to maximize the energy efficiency.

Hence, the joint bandwidth and power allocation approach represents a cross-layer design in a cooperative networking setting.

In such a networking environment, to develop an energy-efficient joint bandwidth and power allocation approach, several challenging issues need to be addressed, as discussed in the next section.

3.3   Challenging Issues

There are various technical challenges towards developing a joint bandwidth and power allocation approach. Further studies are required to deal with single-user versus multiuser systems, single-operator versus multioperator systems, fairness, centralized versus decentralized architectures, number of MT radio interfaces versus number of available networks, in-device coexistence (IDC) interference and computational complexity.

3.3.1   Single-User versus Multiuser System

In the literature, several works investigating energy efficiency of MTs focus on a single-user scenario, for example, [16, 24, 29]. Hence, given an MT that is allocated a fixed amount of bandwidth, the main objective is to allocate the uplink transmission power to maximize the MT energy efficiency. The system model under consideration does not capture the effect of interference among different MTs operating in the same frequency band [46]. Also, a single-user system does not capture the competition among different MTs over the radio resources (e.g. bandwidth [28]) to satisfy their target QoS at an improved energy efficiency.

Joint bandwidth and power allocation calls for a multiuser system to model the competition among different MTs over the available bandwidth at the BSs/APs of different networks, given their different channel conditions and path losses. For instance, in Figure 3.2, MTs 1 and 2 (3 and 4) compete on the available bandwidth at BS 2 on the uplink (downlink) to enhance their energy efficiency. However, such a multiuser system further complicates the analysis, as will be presented in Chapter 5. Rather than maximizing energy efficiency for a single MT, the objective now is to enhance energy efficiency for a set of MTs, which leads to the fairness issue, to be discussed later in more details.

3.3.2   Single-Operator versus Multioperator System

Most of the existing works dedicated to the heterogeneous wireless communications environment assume a single-network operator. This is evident from how different MTs are treated in the context of radio resource management. In the literature, all MTs have been treated equally by different networks in the heterogeneous networking environment. However, this does not take into account the fact that different MTs are subscribers of different operators. Hence, with multi-homing services, each operator wants to first support its own subscribers and to ensure that they are satisfied with the maximum required QoS, while at the same time support the subscribers of other networks. For instance, in the context of joint bandwidth and power allocation, a network operator may allocate bandwidth to the mobile subscriber of another operator only if this operation does not degrade the energy efficiency of its own subscribers. This calls for service differentiation among network subscribers and users. One way to accomplish such a service differentiation is through a priority mechanism [4]. In this case, every network operator gives a higher priority in allocating its resources to its own subscribers as compared to the other users. However, such a priority mechanism may degrade the overall energy efficiency of the networking environment as it limits the exploitation of the available resources. Further studies are required to assess how to maximize the overall energy efficiency, while at the same time support service differentiation among network subscribers and users in the networking setting.

Other implications for modelling a multioperator system are related to fairness and implementation complexity, as discussed in the following subsections.

3.3.3 Fairness

For energy-efficient uplink communications in a multiuser and multioperator system, a popular problem formulation is to maximize the sum energy efficiency for the users in the system, for example, [28, 29] and [46]. However, this objective provides no fairness guarantee for energy efficiency among different MTs. The sum energy efficiency can be maximized by dramatically improving energy efficiency only for a subset of users. One approach to promote fairness is by maximizing the geometric mean of energy efficiency for all MTs, which introduces proportional fairness among all users [28]. Another notion of fairness is based on a max–min formulation, where the objective is to maximize the energy efficiency performance of the worst MT. An investigation of this objective is presented in Chapter 5.

As for energy-efficient downlink communications, most existing studies make no distinction in the amount of energy saved by different operators, and the main objective is to reduce the total energy consumption in the networking environment [12]. However, this can be satisfied by minimizing the energy consumption of one operator at the expense of increasing energy consumption of another operator. It is desired that green communications result in mutual benefits for all operators. The degree of cooperation among different operators is determined based on the attained utility by each operator. The general problem of how to motivate various operators to cooperate among each other to enhance the overall energy efficiency in the networking environment while maximizing the achieved utility for each operator will be investigated in Chapter 4.

3.3.4 Centralized versus Decentralized Implementation

Joint bandwidth and power allocation can be implemented in a centralized or decentralized architecture, as shown in Figure 3.3. In the centralized implementation, a central resource manager is responsible for the resource allocation decisions. The central resource manager has global information of the available resources (e.g. radio bandwidth and maximum transmission power) at BSs/APs of different networks. Also, the central resource manager has information about the required QoS of all MTs in service, the available energy of each MT for uplink communications, and the channel conditions among the MTs and BSs/APs of different networks. Given the collected information, the central resource manager allocates resources from different networks so as to maximize the energy efficiency in the overall networking environment and satisfy the required QoS for all MTs. However, the central resource manager is

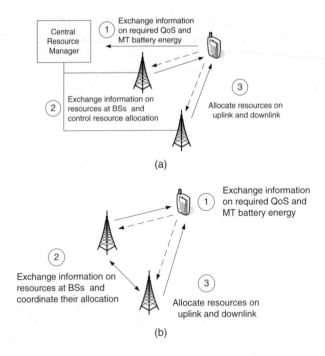

Figure 3.3   Centralized and decentralized implementations [126]. (a) Centralized; (b) decentralized

not a practical solution when different networks are operated by different service providers [4]. A central resource manager who controls the operation of BSs/APs of different networks raises some concerns: (i) the central resource manager creates a single point of failure, hence, if it fails, the whole multi-homing service fails; (ii) the operator in charge of the operation and maintenance of the central resource manager will control the resources of other networks and (iii) changes are required in the structure and operation of different networks to account for the central resource manager.

Given the aforementioned concerns, a decentralized resource management is a more practical and flexible solution. In a decentralized implementation, BSs/APs of different networks make their resource allocation decisions based on their available local information. Better resource allocation decisions can be made through information exchange among BSs/APs of different networks on one side and between BSs/APs and MTs on the other side. However, high signalling overhead is expected in large systems with multiusers and multioperators. Further investigations on how to reduce the associated signalling overhead and guarantee the decentralized solution convergence are presented in Chapter 4 for the green downlink framework and in Chapter 5 for the green uplink communication scenario.

3.3.5   In-device Coexistence Interference

The MT multiple radio interfaces are located in close proximity within the same device. When two radio technologies having adjacent bands operate simultaneously in the same device, the

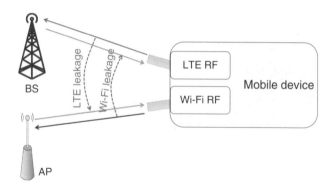

Figure 3.4 Illustration of IDC interference [127]

out-of-band (OOB) radiations of the transmitting radio leak into the band of the receiving radio due to the filter non-idealities. These OOB radiations are commonly referred to as IDC interference [127]. One example of using adjacent frequency bands in a multi-homing scenario exists between the LTE band 40 (B40) and WiFi [128, 129]. LTE B40 ranges between 2,300 and 2,400 MHz, while WiFi operates in the industrial, scientific and medical (ISM) band from 2,401 to 2,500 MHz. For the same MT, the data transmission and reception over LTE occur in a band adjacent to the data transmission and reception over the WiFi. Hence, the downlink reception in each radio suffers from an IDC interference due to the uplink transmission in the other radio, as depicted in Figure 3.4. For video calls supported by the multi-homing capability, the MT can utilize one radio interface (LTE) for the uplink stream and the second radio interface (WLAN) for the downlink stream, and hence, the downlink reception on the WLAN can suffer from IDC interference due to the uplink transmission on the LTE. It has been demonstrated that the currently used filter technology is not sufficient to suppress the IDC interference [128]. Hence, multi-homing radio resource management mechanisms should model such interference and account for it in their resource allocation strategies. However, the existing research overlooks this important aspect. Next, we will discuss how to model such an IDC interference [127].

Every filter has a roll-off factor which defines the amount of frequency band secured for the transition between the pass band and stop band. The IDC interference is caused by these filter non-idealities. The non-ideal steepness of the aggressor WLAN spectrum mask allows part of its power spectral density to leak into its neighbouring LTE band. In addition, the non-ideal selectivity of the victim LTE receiver filter captures the emissions from the WLAN band. This leaked power constitutes the IDC interference. As the frequency separation between the LTE and WLAN channels allocated to the same MT decreases, the IDC interference increases, as depicted in Figure 3.5. An overlap factor (V) can be used to measure the ratio of the leaked WLAN power at the LTE receiver to the total WLAN power [127]. The LTE cellular BS has a set of channels $C = \{1, 2, \ldots, C\}$ that can be assigned to MTs. Each channel $c \in C$ is assigned to a single MT at a time. The WLAN AP has a single channel that is shared by MTs in a TDMA manner using the enhanced version of the Distributed Coordination Function (DCF) [130]. Let $S(f - f_c)$ represent the LTE receiver filter transfer function operating at channel c with frequency f_c. Also, let $S'(f - f')$ denote the WLAN power spectral density centred at f', which is approximated by its transmit spectrum mask. This is a reasonable assumption since

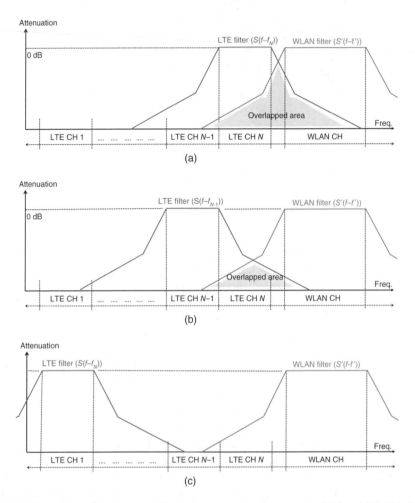

Figure 3.5 Illustration of the impact of frequency separation between the LTE and WLAN channels on the IDC interference [127]. (a) Maximum interference occur for adjacent LTE and WLAN channels; (b) interference decreases as the frequency separation increases between the LTE and WLAN channels; (c) zero interference for sufficiently faraway LTE and WLAN channels

the transmit spectrum mask represents an upper bound of the power spectral density [131]. The overlap factor of the WLAN and LTE filters of MT m, given that the LTE radio is assigned to channel c, is expressed as

$$V_{c,m} = \frac{\int S_m(f - f_c)S'_m(f - f')\mathrm{d}f}{\int S'_m(f - f')\mathrm{d}f}. \tag{3.1}$$

Let P_m^{UL} and A_m denote the interfering WLAN uplink power allocated by MT m and the losses due to the antenna isolation at MT m, respectively. The antenna isolation depends on the physical separation between the WLAN and LTE antennas. The IDC interference that affects

LTE channel c when it is allocated to user m is expressed as

$$I_{c,m} = \frac{P_m^{UL} V_{c,m}}{A_m}.$$ (3.2)

The WLAN reception is also affected by the LTE transmission in the same MT. Under the typical case where each radio has similar transmitter and receiver filters, the overlap factor of interference at the WLAN receiver takes the form

$$V'_{c,m} = \frac{\int S_m(f - f_c) S'_m(f - f') df}{\int S_m(f - f_c) df}.$$ (3.3)

However, the user may be assigned more than one LTE channel. All these channels contribute to the IDC interference at the WLAN receiver I'_m, which is given by

$$I'_m = \sum_{c=1}^{C} x_{c,m} \frac{P_{c,m}^{UL'} V'_{c,m}}{A_m},$$ (3.4)

where $P_{c,m}^{UL'}$ and $x_{c,m}$ denote the interfering LTE uplink power allocated by MT m on channel c and the channel assignment variable that indicates if MT m is assigned to LTE channel c ($x_{c,m} = 1$) or not ($x_{c,m} = 0$), respectively.

In order to show the level of IDC interference that the LTE and WLAN might suffer from, consider the simultaneous operation of LTE and WLAN radio technologies in one mobile, where the WLAN operates on a 22-MHz time division duplex (TDD) channel, while the LTE operates on a 5-MHz TDD channel. The design of the WLAN spectrum mask is obtained from the IEEE 802.11 standard [131]. The used LTE spectrum mask is similar to a 5-MHz ISM filter, but shifted to create a pass band for the LTE channels, as carried out by the 3GPP [128]. Interference parameters are summarized in Table 3.1. Figure 3.6 depicts the interference on the WLAN channels when the aggressor LTE is centred at 2,397.5, 2,387.5 and 2,377.5 MHz. The interference is compared with a typical noise floor in WLAN receivers. As expected, the WLAN channels close to the aggressor LTE suffer from very high interference. The severity of interference across the WLAN channels decreases as we move away from the LTE band until it falls beneath the noise floor. Similarly, the interference caused by the WLAN operation on the LTE channels is depicted in Figure 3.7. The behaviour of the IDC interference on the LTE channels is similar to the WLAN case. The IDC interference can reach undesirable values due to the fact that the interfering and victim antennas lie in close vicinity in one device.

Such an IDC model in the context of a green multi-homing radio resource allocation framework is further investigated in Chapter 4.

Table 3.1 Interference parameters [127]

Parameter	Value
Antenna isolation (A_m)	20 dB
LTE uplink power ($P_{c,m}^{UL}$)	23 dBm
WLAN uplink power (P_m^{UL})	20 dBm

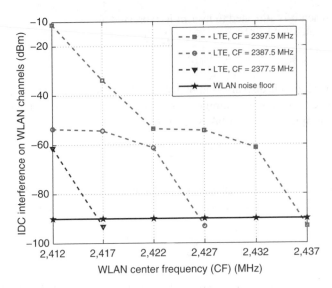

Figure 3.6 IDC interference on different WLAN channels due to the uplink transmission of LTE at 2,397.5, 2,387.5 and 2,377.5 MHz [127]

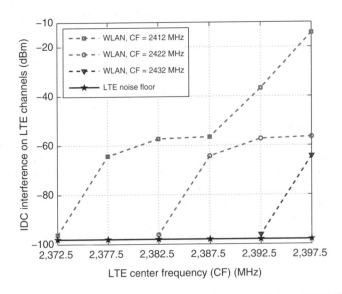

Figure 3.7 IDC interference on LTE channels due to the uplink transmission of WLAN at 2,412, 2,422 and 2,432 MHz [127]

3.3.6 *Computational Complexity*

The joint bandwidth and power allocation approach has a higher computational complexity than a power-only allocation scheme. From a resource allocation perspective, the increased computational complexity is mainly due to the increased number of decision variables (bandwidth and power allocation for all radio interfaces in the joint approach versus only power allocation for all radio interfaces) and constraints. In particular, in uplink resource allocation, the joint bandwidth and power allocation should include the network-side constraints (e.g. the total allocated bandwidth satisfying the BS/AP transmission capacity) in the problem formulation, whereas the power-only allocation mainly deals with the MT-side constraints in terms of the MT available energy and required QoS.

For computational efficiency, the joint bandwidth and power allocation problem can be decomposed into two steps, corresponding to the network-side and MT-side, respectively [54]. In the first step, bandwidth is allocated from the BSs/APs of different networks, on the uplink/downlink, to the MTs, given the required QoS of MTs. This step mainly focuses on the network-side resource allocation (bandwidth) and constraints while satisfying the required QoS, and it can be implemented in the network layer. On the basis of the allocated bandwidth, power allocation is performed in the second step to enhance energy efficiency of the MTs in the uplink and the BSs/APs in the downlink. For uplink resource allocation, this step mainly focuses on the MT-side constraints (MT total available energy) and can be implemented in the physical layer. The resource allocation approach iteratively executes these two steps until energy efficiency is maximized, a strategy that will be discussed in Chapters 4 and 5.

3.3.7 Number of MT Radio Interfaces versus Number of Available Networks

In heterogeneous wireless networks, researchers mainly assume that an MT connects to all existing networks in a multi-homing manner (i.e. the number of MT radio interfaces is equal to the number of available BSs/APs from different networks). This assumption reduces the computational complexity of the problem from a resource allocation perspective as we deal with a nonlinear program (rather than dealing with a mixed-integer nonlinear program (MINLP) as in the case with unequal number of radio interfaces and available BSs/APs). However, such a vision overlooks the fact that the MT is equipped with only a limited number of radio interfaces for each network type, for example, one cellular and one WiFi radio interface, as shown in Figure 3.8. Hence, the MT has to select one cellular BS and one WiFi AP from all the available ones to get its required radio resources in a multi-homing manner. In order to account for

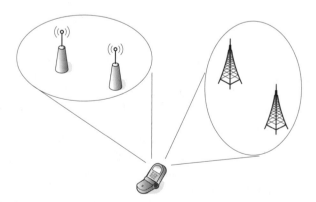

Figure 3.8 The presence of multiple BSs/APs for a limited number of radio interfaces per MT

the limited number of interfaces for MTs, a joint user assignment and bandwidth and power allocation framework is required to support MTs with multi-homing capabilities.

The work in [132] has investigated this challenge in the context of network capacity maximization. Consider a heterogeneous wireless medium that is composed of two networks, namely cellular network and WLAN, that is, $\mathcal{N} = \{1, 2\}$, where 1 represents the cellular network and 2 represents the WLAN. The set of BSs of the cellular network is denoted by $\mathcal{S}_1 = \{1, 2, \ldots, S_1\}$ and the set of APs for the WLAN is represented by $\mathcal{S}_2 = \{S_1 + 1, S_1 + 2, \ldots, S_1 + S_2\}$. The set of available BSs and APs is denoted by $\mathcal{S} = \mathcal{S}_1 \cup \mathcal{S}_2$, where $\mathcal{S}_1 \cap \mathcal{S}_2 = \phi$. Each MT, $m \in \mathcal{M} = \{1, 2, \ldots, M\}$ is equipped with cellular and WLAN radio interfaces, and $\mathcal{L}_1$ and $\mathcal{L}_2$ denote the set of cellular and WLAN radio interfaces, respectively. MT $m \in \mathcal{M}$ asks for a best effort service. Although some of smartphones with dual SIM card are equipped with more than one cellular radio interface, these interfaces cannot be used simultaneously to run two calls, as all cellular interfaces share the same transceiver. Therefore, the MT can have multiple radio interfaces of the same type, but the MT can only utilize one radio interface of one type at a given moment of time. Consequently, the MT can utilize one radio interface from $\mathcal{L}_1$ and one radio interface from $\mathcal{L}_2$ and can aggregate the offered bandwidth from these two radio interfaces to support its ongoing call. Hence, the problem reduces to specify: (i) a binary user assignment decision variable x_{sml} that indicates if MT m can connect to BS/AP s using its radio interface l ($x_{sml} = 1$) or not ($x_{sml} = 0$), (ii) a bandwidth allocation decision variable b_{sml} that indicates the amount of bandwidth allocated to MT m from BS/AP s via radio interface l and (iii) a power allocation decision variable P_{sml} that indicates the amount of power allocated (in the uplink or downlink) on the link connecting MT m to BS/AP s via radio interface l. The binary decision variable constraints (e.g. (3.5) and (3.6)) ensure that an MT connects to a cellular BS using only a (single) cellular radio interface, that is,

$$\sum_{s \in \mathcal{S}_1} \sum_{l \in \mathcal{L}_1} x_{sml} \leq 1, \quad \forall m \in \mathcal{M}, \tag{3.5}$$

$$\sum_{s \in \mathcal{S}_1} \sum_{l \in \mathcal{L}_2} x_{sml} = 0, \quad \forall m \in \mathcal{M}, \tag{3.6}$$

$$\sum_{s \in \mathcal{S}_2} \sum_{l \in \mathcal{L}_2} x_{sml} \leq 1, \quad \forall m \in \mathcal{M}, \tag{3.7}$$

$$\sum_{s \in \mathcal{S}_2} \sum_{l \in \mathcal{L}_1} x_{sml} = 0, \quad \forall m \in \mathcal{M}. \tag{3.8}$$

Similarly, the constraints (3.7) and (3.8) guarantee that an MT connects to a WLAN AP using only a (single) WLAN radio interface. In addition to the aforementioned constraints, the following constraints should be accounted for: (i) capacity limitation of BSs/APs, which accounts for the limitation on the available bandwidth at the BS/AP and (ii) total power consumption constraint (at MT for uplink radio resource allocation or at BS/AP for downlink radio resource allocation). For energy efficiency maximization, the objective function can be expressed by (1.27) for the network operators (i.e. in the downlink radio resource allocation) and (1.28)–(1.32) for the mobile users (i.e. in the uplink radio resource allocation).

The radio resource allocation problem described above is a fractional MINLP due to the binary nature of the MT assignment problem and the continuous nature of the radio resource (bandwidth and power) allocation problem. Hence, higher computational complexity is expected than the approach discussed in the previous subsection. A possible solution relies on exhaustive search, where the MT tries all combinations of cellular network BS and WLAN AP and selects the pair that maximizes the energy efficiency while satisfying the required QoS. However, in a multiuser multioperator system, such an approach incurs high computational complexity and requires high signalling overhead in a decentralized architecture. Furthermore, the heuristic randomization approach proposed in [132] is no longer valid for energy efficiency maximization. The heuristic approach of [132] randomly assigns MTs to BSs and APs and equally allocates the bandwidth available at each BS/AP among the assigned users. However, for energy efficiency maximization, the MT locations and distances from different BSs/APs should be accounted for, which makes the randomization approach not useful in this context. Hence, more efficient techniques should be developed to properly select BSs and APs and allocate the required radio resources.

3.4   Summary

The heterogeneous wireless access medium exhibits great potential in improving energy efficiency while satisfying the QoS of mobile users. Multi-homing services can aggregate bandwidth from different networks, enable better mobility support and reduce energy consumption for mobile users and network operators. In addition to exploiting different channel conditions and path losses among MTs and BSs/APs of different networks, the available bandwidth and operating frequency bands at different networks can further enhance energy efficiency. Hence, a joint bandwidth and power allocation multi-homing approach results in a significant advantage in energy-efficient communications over the power-only allocation scheme. In joint bandwidth and power allocation for uplink and downlink communications in a heterogeneous networking setting, there are many challenging technical issues that require further studies, including fairness in energy efficiency among MTs, achieving mutual benefits among network operators, decentralized implementation with reduced signalling overhead, IDC interference management and implementation complexity. The subsequent chapters are dedicated to address these challenges. Particularly, Chapter 4 investigates achieving mutual power saving among network operators and handling IDC interference management in RF-downlink multi-homing radio resource allocation. Chapter 5 deals with fairness in energy efficiency among MTs, supporting energy-efficient video-streaming applications, and decentralized implementation with reduced signalling overhead and lower computational complexity in RF-uplink multi-homing radio resource allocation. Finally, Chapter 6 studies the adoption of VLC systems in enhancing energy efficiency in the heterogeneous wireless medium via integration with RF networks.

4

Multi-homing for a Green Downlink

In a heterogeneous wireless medium, downlink multi-homing radio resource allocation can save power for network operators. In this context, two application scenarios can be distinguished. In the first scenario, the MT aggregates the offered radio resources from different networks to support a single (data hungry) application, while in the second scenario, the MT runs different applications using the radio resources assigned to different radio interfaces. In this chapter, we discuss the challenging issues associated with each application scenario. Two radio resource allocation mechanisms are presented to address the corresponding challenges.

4.1 Introduction

The increasing demand for wireless communication services and the wide deployment of wireless communication infrastructures have led to high power consumption in the wireless access networks. Due to financial and environmental concerns, wireless network service providers have shown a growing interest in deploying power-efficient wireless communication infrastructures. Given today's heterogeneous wireless medium with overlapped coverage from different networks and the multi-homing capabilities of MTs, green multi-homing radio resource allocation in the downlink can lead to power saving for the network operators. Green multi-homing techniques can be adopted in two application scenarios. The first scenario aggregates the offered radio resources from different networks to support in the downlink a single (data or video) call using multiple threads at the application layer. The second scenario assumes different applications running at the MT, and it can be supported using different radio resources from different networks. For instance, the MT can make a video call using the radio resources offered to one radio interface while performing a data call using the radio resources offered at another radio interface.

Green Heterogeneous Wireless Networks, First Edition. Muhammad Ismail, Muhammad Zeeshan Shakir, Khalid A. Qaraqe and Erchin Serpedin.
© 2016 John Wiley & Sons, Ltd. Published 2016 by John Wiley & Sons, Ltd.

For the first application scenario, several networks with overlapped coverage can cooperate in allocating their radio resources to MTs to satisfy the required service quality at reduced power consumption. Network cooperative (multi-homing) radio resource allocation can lead to a higher power saving than the single-network (non-cooperative) mechanisms, due to the diversity in the wireless channels between MTs and different BSs and in the available resources at the BSs of different networks. However, the existing cooperative networking mechanisms suffer from several limitations. First, there exists an implicit assumption that different networks are willing to cooperate unconditionally to reduce the total power consumption within a certain geographical region (e.g. [32, 33] and [36]). This assumption is valid in the case that all networks are operated by the same service provider. In a cooperative networking scenario, the total (sum) power consumption can be reduced at the cost of increasing the power consumption for one operator, as compared with the non-cooperative case. Hence, when different networks are operated by different service providers, cooperation is appropriate only if it results in mutual benefit (i.e. power saving) for all the networks. The second limitation with existing cooperative networking mechanisms is that they adopt only power allocation techniques, assuming some fixed bandwidth allocation (e.g. [33] and [133]). Such an approach focuses mainly on exploiting only one dimension in the heterogeneous wireless medium, which is the diverse fading channels and propagation losses among MTs and BSs of different networks. Another dimension that should be accounted for is the disparity in the available bandwidth at BSs of different networks. Therefore, joint bandwidth and power allocation (within the framework described in Chapter 3) should be investigated for improved power saving in such a networking environment. Finally, the existing mechanisms rely on a central manager for resource allocation (e.g. [134] and references therein). Such mechanisms are useful in a single-operator scenario; however, they are infeasible in a multi-operator scenario due to the absence of a central entity. Instead, decentralized mechanisms that enable coordinated resource allocation among different networks are required.

On the other hand, for the second application scenario, when two radio interfaces operate simultaneously using different radio technologies with adjacent frequency bands (e.g. LTE and WiFi), the OOB radiations of the transmitting radio (on the uplink) leak on the receiving radio (on the downlink), and hence, the MT suffers from IDC interference as discussed in Chapter 3. Without appropriate consideration of the IDC interference in the radio resource allocation, network operators will be subject to high power consumption in the downlink to overcome the interference of the MT uplink transmission of the other radio interface and satisfy the mobile user's required QoS. Consequently, power wastage is expected in the downlink. However, none of the existing research has modelled the IDC interference or accounted for it in the resource allocation problem [135, 136].

In this chapter, we present two downlink radio resource allocation mechanisms that address the aforementioned research issues for the two application scenarios. In particular, for the first application scenario, we present a decentralized downlink joint radio resource (bandwidth and power) allocation strategy that guarantees a win–win situation among different service providers so that they have incentive to cooperate for power saving [137]. For the second application scenario, we present a downlink radio resource allocation mechanism that considers the IDC interference in LTE and WLAN, and meets the data rate requirements in the victim downlink path by using an intelligent scheme for joint channel, time and downlink power radio resource allocation [127].

4.2   Win–Win Cooperative Green Resource Allocation

In this section, the downlink green communication problem is formulated as an asymmetric
Nash bargain game to jointly allocate radio resources (bandwidth and power) from different networks to a set of MTs with multi-homing capabilities. The game formulation captures
bargain powers of different networks based on each network capability (e.g. total available
bandwidth). We show that there exist a bargain solution and unique bargain point for the Nash
bargain game, which ensure that the allocated radio resources result in power saving for different service providers as compared with the non-cooperative scenario, and hence such a strategy
motivates them to cooperate. We present a decentralized solution for the Nash bargain game,
where MTs coordinate the allocation from different networks to satisfy the required QoS at BS
with reduced power consumption. Such a decentralized solution is desirable in a multi-operator
system in the absence of a central resource manager.

Consider a geographical region covered by a set $\mathcal{N} = \{1, 2, \cdots, N\}$ of cellular networks,
as shown in Figure 4.1. The cellular networks are operated in separate frequency bands
by different service providers, and hence, no interference exists among these networks.
Each cellular network has a set of BSs covering the geographical region, denoted by
$\mathcal{S}_n = \{1, 2, \cdots, S_n\}$. Interference management techniques (e.g. soft frequency reuse
[138–140] and [141]) are adopted for interference mitigation among BSs of the same
network. As the BSs of different networks present overlapped coverage, the geographical
region is partitioned into a set of service areas. A unique subset of BSs covers each service
area. The BS of network n has a static (fixed) power consumption $P_{ns,f}$, independent of
the call traffic load, which represents the power consumed in the cooling, backhaul and
other circuits. The power amplifier efficiency for each BS is denoted by ξ. The total power

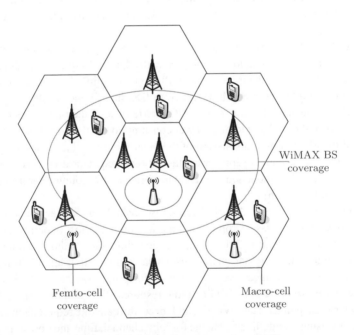

Figure 4.1   Network coverage areas [137]

consumption of network n BS s has a maximum constraint of $P_{ns,\max}$. The maximum available bandwidth at network n BS s is $B_{ns,\max}$.

Let $\mathcal{M}_{ns}$ denote the set of network n active subscribers in the coverage area of BS $s \in \mathcal{S}_n$. The set of active MTs in the geographical region is denoted by $\mathcal{M} = \cup_{n,s} \mathcal{M}_{ns}$. MTs are equipped with multiple radio interfaces and multi-homing capabilities. Let P_{nsm} and B_{nsm} denote the power and bandwidth allocated from network $n \in \mathcal{N}$ BS $s \in \mathcal{S}_n$ to MT $m \in \mathcal{M}$, respectively.[1] In a non-cooperative (single-network) scenario, each MT m obtains the required radio resources to satisfy a target minimum data rate $R_{m,\min}$ from its home network. In a cooperative networking scenario, an MT obtains the required radio resources from all networks available at its location. In this section, we do not include a priority mechanism in the problem formulation to differentiate network subscribers and other users when cooperative networking is employed. Hence, all MTs are treated equally by all operators in the cooperative case. Also, it is assumed that the number of radio interfaces for each MT is the same as the number of available BSs/APs at its location.

The channel power gain between network n BS s and MT m is denoted by h_{nsm}, and it captures both the path loss and fast fading. The path loss is given by $d_{nsm}^{-\alpha}$, which is proportional to the distance between network n BS s and MT m and α denotes the path-loss exponent. The one-sided noise power spectral density is denoted by N_0.

4.2.1 Non-cooperative Single-Network Solution

In this case, different networks do not cooperate with each other for power saving. Hence, each network n BS s allocates its radio resources only to its subscribers, that is, for $m \in \mathcal{M}_{ns}$, as shown in Figure 4.2a. The objective is to minimize the power consumption of network n while satisfying the required minimum QoS for the network subscribers and the BSs' radio resource constraints. This can be expressed by

$$\min_{\{P_{nsm}^{(NCS)}, B_{nsm}^{(NCS)}\} \geq 0} P_n^{(NCS)}$$

$$\text{s.t.} \quad R_{nsm}^{(NCS)} \geq R_{m,\min}, \quad \forall m \in \mathcal{M}_{ns}, s \in \mathcal{S}_n$$

$$B_{ns}^{(NCS)} \leq B_{ns,\max}, \quad \forall s \in \mathcal{S}_n$$

$$P_{ns}^{(NCS)} \leq P_{ns,\max}, \quad \forall s \in \mathcal{S}_n, \tag{4.1}$$

where NCS denotes the non-cooperative solution. The total power consumption for network n, $P_n^{(NCS)}$, is expressed as the summation of the power consumption of the BSs of this network, that is, $P_n^{(NCS)} = \sum_{s \in \mathcal{S}_n} P_{ns}^{(NCS)}$, and $P_{ns}^{(NCS)}$ consists of two components, namely the load-independent and load-dependent components, that is,

$$P_{ns}^{(NCS)} = P_{ns,f} + \xi \sum_{m \in \mathcal{M}_{ns}} P_{nsm}^{(NCS)}. \tag{4.2}$$

---

[1] Since in this section we focus only on a downlink scenario, we omit the superscripts UL and DL from all mathematical symbols, unlike our definitions in Chapter 3.

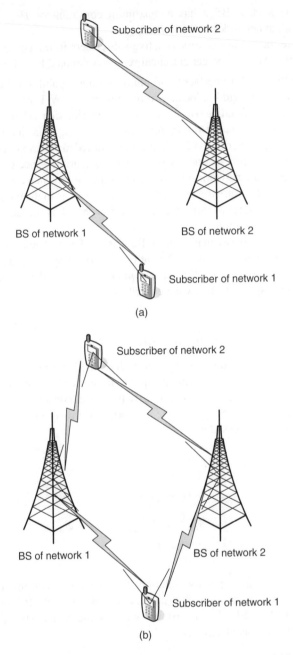

Figure 4.2   Illustration of non-cooperative single-network and cooperative multi-homing radio resource allocation. (a) Non-cooperative single-network solution; (b) Cooperative multi-homing solution

The achieved data rate for MT m on the downlink of BS s of home network n is given by Shannon formula

$$R_{nsm}^{(NCS)} = B_{nsm}^{(NCS)} \log_2 \left(1 + \frac{P_{nsm}^{(NCS)} h_{nsm}}{N_0 B_{nsm}^{(NCS)}} \right). \qquad (4.3)$$

The total bandwidth allocated by network n BS s to its subscribers is given by $B_{ns}^{(NCS)} = \sum_{m \in \mathcal{M}_{ns}} B_{nsm}^{(NCS)}$.

Since the Hessian matrix of R_{nsm} is negative semi-definite [137], R_{nsm} is concave in B_{nsm} and P_{nsm}. As a result, (4.1) minimizes a linear objective function over a convex set (as the first constraint in (4.1) is convex in P_{nsm} and B_{nsm} and the second and third constraints are linear in P_{nsm} and B_{nsm}). Thus, (4.1) is a convex optimization problem [142].

Since (4.1) is a convex optimization problem, it can be solved efficiently by each network BS in polynomial time complexity to find a green downlink radio resource allocation (i.e. $P_{nsm}^{(NCS)}$ and $B_{nsm}^{(NCS)}$) for the network subscribers.

4.2.2 Win–Win Cooperative Solution

In a heterogeneous networking environment, networks can cooperate in radio resource allocation to achieve power saving, as shown Figure 4.2b. Using the multi-homing capability, MTs aggregate the offered radio resources (i.e. bandwidth and power) from the BSs of different networks with overlapped coverage to satisfy the required minimum QoS and the bandwidth and power constraints of different BSs, to yield power saving for service providers. However, in a multi-operator system, where BSs with overlapped coverage are operated by different service providers, a rationale service provider prefers to work independently (similar to the previous subsection) if its power consumption through cooperation is higher than the NCS. Consequently, in order to promote cooperation for power saving among service providers, Nash bargain games [143] can be used to offer a cooperation incentive radio resource allocation that guarantees mutual benefit (power saving) among different service providers. In this case, the players of the game are the service providers, which are given by the set $\mathcal{N}$. Define Ω^N as the set of game strategies of the N networks, that is, $(P_{nsm}^{(NBS)}, B_{nsm}^{(NBS)}) \; \forall n \in \mathcal{N}, s \in \mathcal{S}_n, m \in \mathcal{M}$, where NBS denotes the Nash bargain solution and $P_{nsm}^{(NBS)} = B_{nsm}^{(NBS)} = 0$ if m is not in the coverage area of network n BS s. Let Υ be the space of the radio resource allocation strategies, expressed by

$$\Upsilon = \{\{P_{nsm}^{(NBS)}, B_{nsm}^{(NBS)} \; \forall n, s, m \mid \sum_{n \in \mathcal{N}} \sum_{s \in \mathcal{S}_n} R_{nsm}^{(NBS)} \geq R_{m,\min} \; \forall m,$$

$$B_{ns}^{(NBS)} \leq B_{ns,\max} \; \forall s, P_{ns}^{(NBS)} \leq P_{ns,\max} \; \forall s\}, \qquad (4.4)$$

where $R_{nsm}^{(NBS)}$ is obtained by replacing NCS by NBS in (4.3), $B_{ns}^{(NBS)} = \sum_m B_{nsm}^{(NBS)}$, and

$$P_{ns}^{(NBS)} = P_{ns,f} + \xi \sum_m P_{nsm}^{(NBS)}. \qquad (4.5)$$

Assuming (4.1) is feasible, the constraint set (4.1) forms a sub-space of Υ. This can be easily shown by observing that the constraints in (4.1) can be written as the constraints in (4.4) with $P_{nsm} = B_{nsm} = 0 \; \forall m \notin \mathcal{M}_{ns}$. Hence, the constraint set (4.1) is a special case of Υ.

We aim to find a feasible sub-space $\Upsilon^* \in \Upsilon$ with radio resource allocation that results in total power consumption $P_n^{(\text{NBS})}$ less than or equal to $P_n^{(\text{NCS})} \ \forall n \in \mathcal{N}$. In Nash bargain game theory, $P_n^{(\text{NCS})}$ is referred to as the *initial agreement point*. Define $\mathcal{P} = \{(P_n, P_n^{(\text{NCS})}) \forall n \in \mathcal{N} | \Omega^N \in \Upsilon\}$ as a set of network power consumptions satisfying the initial agreement point for $n \in \mathcal{N}$. A mapping $\mathcal{F} : \mathcal{P} \rightarrow \Omega^N$ is a symmetric NBS if it satisfies the following axioms [143]:

1. $\mathcal{F}(P_n, P_n^{(\text{NCS})})$ is Pareto optimal, that is, there exists no other radio resource allocation that results in a superior performance for all networks simultaneously.
2. $\mathcal{F}(P_n, P_n^{(\text{NCS})})$ is independent of irrelevant alternatives, that is, if the feasible set shrinks but the radio resource allocation remains feasible, the radio resource allocation for the smaller feasible set will be the same.
3. $\mathcal{F}(P_n, P_n^{(\text{NCS})})$ is invariant with respect to affine transformations, that is, if $P_n^{(\text{NBS})}$ is the solution of $\mathcal{F}(P_n, P_n^{(\text{NCS})})$ and θ is a positive linear transformation, $\theta(P_n^{(\text{NBS})})$ is the solution of $\theta(\mathcal{F}(P_n, P_n^{(\text{NCS})}))$.
4. $\mathcal{F}(P_n, P_n^{(\text{NCS})})$ provides symmetry, that is, networks with same power consumption in the non-cooperative solution achieve same power saving in the cooperative case.

An asymmetric NBS [144] enables different networks to have different bargain powers in the game, which eventually affects the power saving outcomes for each network. Such a model is very useful to represent the resource availability at different operators (i.e. their capabilities). For instance, one operator may have high network capacity and hence should be able to have more influence on the game outcome.

Let $\widetilde{\mathcal{N}}$ be the non-empty subset of networks that can achieve strictly superior performance through cooperation. There exist a bargaining solution and unique bargain point for the win–win cooperative downlink green radio resource allocation problem, which is obtained by solving

$$\max_{\{P_{nsm}^{(\text{NBS})}, B_{nsm}^{(\text{NBS})}\} \geq 0} \prod_{\widetilde{\mathcal{N}}} (P_n^{(\text{NCS})} - P_n^{(\text{NBS})})^{\varrho_n}$$

$$\text{s.t.} \ \ P_{nsm}^{(\text{NBS})}, B_{nsm}^{(\text{NBS})} \in \Upsilon, \quad \forall \widetilde{\mathcal{N}}, \mathcal{S}_n, \bigcup_{\widetilde{\mathcal{N}}, \mathcal{S}_n} \mathcal{M}_{ns},$$

$$P_n^{(\text{NBS})} < P_n^{(\text{NCS})}, \quad \forall n \in \widetilde{\mathcal{N}} \tag{4.6}$$

where ϱ_n denotes the bargain power of network $n \in \widetilde{\mathcal{N}}$.

In order to prove the existence of a unique bargain point for the win–win cooperative solution, we first prove that Υ is compact and convex. From (4.4), Υ is closed and bounded, and hence it is compact. As R_{nsm} is concave in B_{nsm} and P_{nsm}, $\sum_{n \in \widetilde{\mathcal{N}}} \sum_{s \in \mathcal{S}_n} R_{nsm}^{(\text{NBS})}$ is a sum of concave functions in $P_{nsm}^{(\text{NBS})}$ and $B_{nsm}^{(\text{NBS})}$. Hence, the first constraint in (4.4) is convex. As Υ is described by convex and linear constraints, it is convex. From Theorem 2.1 in [145], since $P_n^{(\text{NBS})} \ \forall n \in \mathcal{N}$ are linear functions defined on the convex and compact set Υ, the cooperative downlink green radio resource allocation has a unique bargain point, which is obtained by solving (4.6).

The optimization problem (4.6) is solved for $n \in \widetilde{\mathcal{N}}$. Networks $n \notin \widetilde{\mathcal{N}}$ follow the NCS. One way to reflect the bargain power ϱ_n of network $n \in \widetilde{\mathcal{N}}$ in (4.6) is based on the ratio of the total bandwidth available at network n to the total bandwidth available in all networks in $\widetilde{\mathcal{N}}$, that is,

$$\varrho_n = \frac{\sum_{s \in \mathcal{S}_n} B_{ns,\max}}{\sum_{n \in \widetilde{\mathcal{N}}} \sum_{s \in \mathcal{S}_n} B_{ns,\max}}. \tag{4.7}$$

From Theorem 2.2 in [145], as $P_n^{\text{(NBS)}}$ is injective over the constraint set in (4.6) $\forall n \in \widetilde{\mathcal{N}}$, an equivalent form to (4.6) is given by

$$\max_{\{P_{nsm}^{\text{(NBS)}}, B_{nsm}^{\text{(NBS)}}\} \geq 0} \sum_{n \in \widetilde{\mathcal{N}}} \varrho_n \log(P_n^{\text{(NCS)}} - P_n^{\text{(NBS)}})$$

$$\text{s.t. } P_{nsm}^{\text{(NBS)}}, B_{nsm}^{\text{(NBS)}} \in \Upsilon, \quad \forall \widetilde{\mathcal{N}}, \mathcal{S}_n, \underset{\widetilde{\mathcal{N}}, \mathcal{S}_n}{\cup} \mathcal{M}_{ns},$$

$$P_n^{\text{(NBS)}} < P_n^{\text{(NCS)}}, \quad \forall n \in \widetilde{\mathcal{N}}. \tag{4.8}$$

From (4.8), it is evident that the asymmetric Nash bargain game (4.6) promotes weighted proportional fairness in power saving among the networks [144]. With equal bargain powers (i.e. symmetric Nash bargain game), proportional fairness is guaranteed among different networks.

As $P_n^{\text{(NCS)}} - P_n^{\text{(NBS)}}$ is concave and positive, for $n \in \widetilde{\mathcal{N}}$, $\log(P_n^{\text{(NCS)}} - P_n^{\text{(NBS)}})$ is also concave [142]. Hence, the objective function of (4.8) is concave, since it is expressed as the sum of concave functions. In addition, the constraint set of (4.8) is convex. Consequently, (4.8) maximizes a concave function over a convex set, and therefore, it represents a convex optimization program.

The Lagrangian function for (4.8) is given by

$$L(P_{nsm}^{\text{(NBS)}}, B_{nsm}^{\text{(NBS)}}, \nu_m^{\text{(NBS)}}, \lambda_{ns}^{\text{(NBS)}}, \beta_{ns}^{\text{(NBS)}}, \mu_n^{\text{(NBS)}}) =$$

$$\sum_{n \in \widetilde{\mathcal{N}}} \varrho_n \log(P_n^{\text{(NCS)}} - P_n^{\text{(NBS)}}) + \sum_m \nu_m^{\text{(NBS)}} \left(\sum_{n \in \widetilde{\mathcal{N}}} \sum_{s \in \mathcal{S}_n} B_{nsm}^{\text{(NBS)}} \cdot \right.$$

$$\left. \log_2 \left(1 + \frac{P_{nsm}^{\text{(NBS)}} h_{nsm}}{N_0 B_{nsm}^{\text{(NBS)}}} \right) - R_{m,\min} \right) + \sum_{n \in \widetilde{\mathcal{N}}} \sum_{s \in \mathcal{S}_n} \lambda_{ns}^{\text{(NBS)}} \cdot \tag{4.9}$$

$$\left(B_{ns,\max} - \sum_m B_{nsm}^{\text{(NBS)}} \right) + \sum_{n \in \widetilde{\mathcal{N}}} \sum_{s \in \mathcal{S}_n} \beta_{ns}^{\text{(NBS)}} \left(P_{n,\max} - \right.$$

$$\left. \left\{ P_{ns,f} + \xi \sum_m P_{nsm}^{\text{(NBS)}} \right\} \right) + \sum_{n \in \widetilde{\mathcal{N}}} \mu_n^{\text{(NBS)}} (P_n^{\text{(NCS)}} - P_n^{\text{(NBS)}}),$$

where $\nu_m^{\text{(NBS)}}$, $\lambda_{ns}^{\text{(NBS)}}$, $\beta_{ns}^{\text{(NBS)}}$ and $\mu_n^{\text{(NBS)}}$ are the Lagrangian multipliers for the data rate constraint of MT m, maximum bandwidth and power constraints for network n BS s, and power saving constraint for network n, respectively. As (4.8) is a convex optimization problem, strong duality exists and we can solve (4.8) through its dual problem [142]. The dual function is expressed as

$$D(\nu_m^{\text{(NBS)}}, \lambda_{ns}^{\text{(NBS)}}, \beta_{ns}^{\text{(NBS)}}, \mu_n^{\text{(NBS)}}) =$$

$$\max_{\{P_{nsm}^{\text{(NBS)}}, B_{nsm}^{\text{(NBS)}}\} \geq 0} L(P_{nsm}^{\text{(NBS)}}, B_{nsm}^{\text{(NBS)}}, \nu_m^{\text{(NBS)}}, \lambda_{ns}^{\text{(NBS)}}, \beta_{ns}^{\text{(NBS)}}, \mu_n^{\text{(NBS)}}). \tag{4.10}$$

Thus, the dual problem is formulated as

$$\min_{\{\nu_m^{\text{(NBS)}}, \lambda_{ns}^{\text{(NBS)}}, \beta_{ns}^{\text{(NBS)}}, \mu_n^{\text{(NBS)}}\} \geq 0} D(\nu_m^{\text{(NBS)}}, \lambda_{ns}^{\text{(NBS)}}, \beta_{ns}^{\text{(NBS)}}, \mu_n^{\text{(NBS)}}). \tag{4.11}$$

We solve (4.10) and (4.11) to find the optimal joint bandwidth and power allocation for cooperative green communications.

4.2.2.1 Downlink Power Allocation at Each Network

In the following, we derive the optimal allocated power at each network BS, given the bandwidth allocation $B_{nsm}^{\text{(NBS)}}$ and Lagrangian multipliers $\lambda_{ns}^{\text{(NBS)}}$ and $\nu_m^{\text{(NBS)}}$. Applying the Karush–Kuhn–Tucker (KKT) conditions [142], it follows that

$$\frac{\partial L(P_{nsm}^{\text{(NBS)}}, B_{nsm}^{\text{(NBS)}}, \nu_m^{\text{(NBS)}}, \lambda_{ns}^{\text{(NBS)}}, \beta_{ns}^{\text{(NBS)}}, \mu_n^{\text{(NBS)}})}{\partial P_{nsm}^{\text{(NBS)}}} = 0. \tag{4.12}$$

It turns out that

$$P_{nsm}^{\text{(NBS)}} = B_{nsm}^{\text{(NBS)}} \left[\frac{\nu_m^{\text{(NBS)}}}{\log(2)(\xi\{\beta_{ns}^{\text{(NBS)}} + \mu_n^{\text{(NBS)}} + \gamma_n/(P_n^{\text{(NCS)}} - P_n^{\text{(NBS)}})\})} - \frac{N_0}{h_{nsm}} \right]^+, \tag{4.13}$$

where $[\cdot]^+$ is a projection on the positive quadrant to account for $P_{nsm}^{\text{(NBS)}} \geq 0$ and $P_n^{\text{(NCS)}}$ is given by the solution of (4.1).

The optimal dual variable $\beta_{ns}^{\text{(NBS)}}$ guarantees that the total allocated power by each BS satisfies its maximum power constraint and can be found using the gradient descent method as follows

$$\beta_{ns}^{\text{(NBS)}}(i+1) = \left[\beta_{ns}^{\text{(NBS)}}(i) - \delta_1 \left(P_{ns,\max} - \left\{ P_{ns,f} + \xi \sum_m P_{nsm}^{\text{(NBS)}}(i) \right\} \right) \right]^+, \tag{4.14}$$

where i is an iteration index and δ_1 is a small step size.

Similarly, the optimal dual variable $\mu_n^{\text{(NBS)}}$ guarantees that the total power allocation by network n, using the cooperative approach, is smaller than that obtained via the non-cooperative approach. Applying the gradient descent method, we have

$$\mu_n^{\text{(NBS)}}(i+1) = [\mu_n^{\text{(NBS)}}(i) - \delta_2(P_n^{\text{(NCS)}} - P_n^{\text{(NBS)}}(i))]^+, \tag{4.15}$$

where δ_2 is a sufficiently small step size.

It can be observed from (4.13) that the power allocation $P_{nsm}^{\text{(NBS)}}$ is a function of the total power consumption $P_n^{\text{(NBS)}}$, which is also a function of $P_{nsm}^{\text{(NBS)}}$. As a result, $P_{nsm}^{\text{(NBS)}}$ is obtained by updating $P_n^{\text{(NBS)}}$ in an iterative way as follows [146]

$$P_n^{\text{(NBS)}}(j+1) = P_n^{\text{(NBS)}}(j) - \delta_3 \left(P_n^{\text{(NBS)}}(j) - \sum_{s \in \mathcal{S}_n} \left(P_{ns,f} + \xi \sum_m P_{nsm}^{\text{(NBS)}}(j) \right) \right) \tag{4.16}$$

where j is an iteration index and δ_3 is a sufficiently small step size. Algorithm 4.2.1 describes the optimal power allocation at each network, and I and J denote the number of iterations required for convergence.

Algorithm 4.2.1 Power Allocation at Each Network $n \in \widetilde{\mathcal{N}}$

Input: $B_{nsm}^{(\text{NBS})}, \lambda_{ns}^{(\text{NBS})}$, and $\nu_m^{(\text{NBS})}$;

Initialization: $(\beta_{ns}^{(\text{NBS})}(1), \mu_n^{(\text{NBS})}(1)) \geq 0$;

for $i = 1 : I$ **do**

    Initialization: $P_n^{(\text{NBS})}(1) < P_n^{(\text{NCS})}$

    for $j = 1 : J$ **do**

        for $s \in \mathcal{S}_n, m \in \underset{\widetilde{\mathcal{N}}, \mathcal{S}_n}{\cup} \mathcal{M}_{ns}$ **do**

$$P_{nsm}^{(\text{NBS})}(j) = B_{nsm}^{(\text{NBS})} \left[\frac{\nu_m^{(\text{NBS})}}{\log(2)(\xi\{\beta_{ns}^{(\text{NBS})}(i) + \mu_n^{(\text{NBS})}(i) + \varrho_n / (P_n^{(\text{NCS})} - P_n^{(\text{NBS})}(j))\})} - \frac{N_0}{h_{nsm}} \right]^+$$

        end for

$$P_n^{(\text{NBS})}(j+1) = P_n^{(\text{NBS})}(j) - \delta_3(P_n^{(\text{NBS})}(j) - \sum_{s \in \mathcal{S}_n}(P_{ns,\text{f}} + \xi \sum_{m \in \mathcal{M}} P_{nsm}^{(\text{NBS})}(j)))$$

    end for

    $P_{nsm}^{(\text{NBS})}(i) = P_{nsm}^{(\text{NBS})}(J)$;

    for $s \in \mathcal{S}_n$ **do**

$$\beta_{ns}^{(\text{NBS})}(i+1) = [\beta_{ns}^{(\text{NBS})}(i) - \delta_1(P_{n,\max} - \{P_{ns,\text{stat}} + \xi \sum_{m \in \mathcal{M}} P_{nsm}^{(\text{NBS})}(i)\})]^+;$$

    end for

    $\mu_n^{(\text{NBS})}(i+1) = [\mu_n^{(\text{NBS})}(i) - \delta_2(P_n^{(\text{NCS})} - P_n^{(\text{NBS})}(i))]^+;$

end for

Output: $P_{nsm}^{(\text{NBS})} \ \forall s, m \in \underset{\widetilde{\mathcal{N}}, \mathcal{S}_n}{\cup} \mathcal{M}_{ns}.$

4.2.2.2 Downlink Bandwidth Allocation at Each Network $n \in \widetilde{\mathcal{N}}$

In the following, we derive the optimal allocated bandwidth at each network BS, given the calculated power allocation $P_{nsm}^{(\text{NBS})}$ and Lagrangian multiplier $\nu_m^{(\text{NBS})}$. Applying the KKT conditions, we have

$$\frac{\partial L(P_{nsm}^{(\text{NBS})}, B_{nsm}^{(\text{NBS})}, \nu_m^{(\text{NBS})}, \lambda_{ns}^{(\text{NBS})}, \beta_{ns}^{(\text{NBS})}, \mu_n^{(\text{NBS})})}{\partial B_{nsm}^{(\text{NBS})}} = 0. \tag{4.17}$$

Therefore, it follows that

$$\log_2 \left(1 + \frac{P_{nsm}^{(\text{NBS})} h_{nsm}}{N_0 B_{nsm}^{(\text{NBS})}} \right) - \frac{P_{nsm}^{(\text{NBS})} h_{nsm}}{\log(2)(N_0 B_{nsm}^{(\text{NBS})} + P_{nsm}^{(\text{NBS})} h_{nsm})} = \frac{\lambda_{ns}^{(\text{NBS})}}{\nu_m^{(\text{NBS})}}. \tag{4.18}$$

The allocated bandwidth $B_{nsm}^{(\text{NBS})}$ can be found as the positive real root of (4.18) using Newton's method.

The optimal dual variable $\lambda_{ns}^{(\text{NBS})}$ guarantees that the total allocated bandwidth by each BS satisfies its maximum bandwidth constraint and can be found using the gradient descent method as follows

$$\lambda_{ns}^{(\text{NBS})}(i+1) = \left[\lambda_{ns}^{(\text{NBS})}(i) - \delta_4 \left(B_{ns,\max} - \sum_m B_{nsm}^{(\text{NBS})}(i) \right) \right]^+, \tag{4.19}$$

where δ_4 is a sufficiently small step size.

Algorithm 4.2.2 describes the optimal bandwidth allocation at each network.

Algorithm 4.2.2 Bandwidth Allocation at Each Network $n \in \widetilde{N}$

Input: $P_{nsm}^{(NBS)}$ and ν_m;
Initialization: $\lambda_{ns}(1) \geq 0$;
for $i = 1 : I$ **do**
    for $s \in S_n$ **do**
        for $m \in \underset{\widetilde{N}, S_n}{\cup} \mathcal{M}_{ns}$ **do**
           Find $B_{nsm}^{(NBS)}(i)$ as the positive real root of(4.18);
        end for
        $\lambda_{ns}^{(NBS)}(i+1) = [\lambda_{ns}^{(NBS)}(i) - \delta_4(B_{ns,\max} - \sum_m B_{nsm}^{(NBS)}(i))]^+$;
    end for
end for
Output: $B_{nsm}^{(NBS)}$ $\forall s, m \in \underset{\widetilde{N}, S_n}{\cup} \mathcal{M}_{ns}$.

4.2.2.3 Downlink Joint Optimal Bandwidth and Power Allocation

The Lagrangian multiplier ν_m is used by each MT $m \in \underset{\widetilde{N}, S_n}{\cup} \mathcal{M}_{ns}$ to coordinate the radio resources allocated by different networks such that the MT minimum required data rate is satisfied. The Lagrangian multiplier $\nu_m^{(NBS)}$ can be found using a gradient descent method as follows

$$\nu_m^{(NBS)}(i+1) = \left[\nu_m^{(NBS)}(i) - \delta_5 \left(\sum_{n \in \widetilde{N}} \sum_s B_{nsm}^{(NBS)}(i) \log_2 \left(1 + \frac{P_{nsm}^{(NBS)}(i) h_{nsm}}{N_0 B_{nsm}^{(NBS)}(i)} \right) - R_{m,\min} \right) \right]^+,$$
(4.20)

where δ_5 is a sufficiently small step size.

Algorithm 4.2.3 describes the optimal joint bandwidth and power allocation that satisfies the MTs' required QoS and exhibit win–win power saving for the service providers.

Algorithm 4.2.3 Joint Optimal Bandwidth and Power Allocation

Initialization: $\nu_m^{(NBS)}(1) \geq 0$;
for $i = 1 : I$ **do**
    for $n \in \widetilde{N}$ **do**
        Allocate power from BSs of different networks using Algorithm 4.2.1;
        Allocate bandwidth from BSs of different networks using Algorithm 4.2.2;
    end for
    for $m \in \underset{\widetilde{N}, S_n}{\cup} \mathcal{M}_{ns}$ **do**
        $\nu_m^{(NBS)}(i+1) = [\nu_m^{(NBS)}(i) - \eta_5 (\sum_n \sum_s B_{nsm}^{(NBS)}(i) \log_2(1 + \frac{P_{nsm}^{(NBS)}(i) h_{nsm}}{N_0 B_{nsm}^{(NBS)}}(i)) - R_{m,\min})]^+$;
    end for
end for
Output: $B_{nsm}^{(NBS)}$ and $P_{nsm}^{(NBS)}$ $\forall n \in \widetilde{N}, s, m \in \underset{\widetilde{N}, S_n}{\cup} \mathcal{M}_{ns}$.

To summarize, network operators should first solve (4.1) to find the NCS (the initial agreement point). Since (4.1) is a convex optimization problem, it can be solved efficiently by each network BS in polynomial time complexity. Then, networks that belong to $\widetilde{\mathcal{N}}$ solve (4.8) to find the NBS. The set of networks belonging to $\widetilde{\mathcal{N}}$ is the one that makes (4.8) feasible. Determining $\widetilde{\mathcal{N}}$ presents an upper bound complexity on the order of $O(2^N)$. Such a complexity is not high since the number of operators within a region, N, is expected to be in the range $[2, 4]$. The complexity of obtaining the NBS using the Algorithm 4.2.3 reduces to solving the dual problem, which presents a complexity upper bounded by $O(M|\widetilde{\mathcal{N}}|\sum_{n\in\widetilde{\mathcal{N}}}S_n)$, where M and $|\widetilde{\mathcal{N}}|$ are the numbers of MTs in $\mathcal{M}$ and networks in $\widetilde{\mathcal{N}}$. Hence, it has a linear complexity in the number of MTs M, network operators $|\widetilde{\mathcal{N}}|$ and BSs $\sum_{n\in\widetilde{\mathcal{N}}}S_n$. Algorithms 4.2.1–4.2.3 are guaranteed to converge to an optimal solution using sufficiently small step sizes, since (4.8) is convex.

4.2.3   Benchmark: Sum Minimization Solution

In this subsection, we present a benchmark for power-efficient cooperative radio resource allocation in a heterogeneous wireless medium. The benchmark is referred to as a sum minimization solution as it uses cooperation among networks to minimize the total (sum) power consumption in the geographical region. Hence, MTs are allocated radio resources (i.e. bandwidth and power) from the BSs of different networks and aggregate them using the multi-homing capability to satisfy its required QoS while minimizing the networks' total power consumption and satisfying the radio resource constraints. This can be expressed by

$$\min_{\{P_{nsm}^{\text{(SMS)}}, B_{nsm}^{\text{(SMS)}}\}\geq 0} \sum_n P_n^{\text{(SMS)}}$$

$$\text{s.t.} \qquad \sum_n \sum_s R_{nsm}^{\text{(SMS)}} \geq R_{m,\min}, \quad \forall m \in \mathcal{M}$$

$$B_{ns}^{\text{(SMS)}} \leq B_{ns,\max}, \quad \forall s \in \mathcal{S}_n, n \in \mathcal{N}$$

$$P_{ns}^{\text{(SMS)}} \leq P_{n,\max}, \quad \forall s \in \mathcal{S}_n, n \in \mathcal{N},$$

$$(4.21)$$

where SMS denotes the sum minimization solution. Similar to (4.8), (4.21) is a convex optimization problem as it involves the minimization of a linear function over a convex constraint set. Hence, (4.21) can be solved efficiently in polynomial time complexity to find a green downlink multi-homing resource allocation (i.e. $P_{nsm}^{\text{(SMS)}}$ and $B_{nsm}^{\text{(SMS)}}$).

4.2.4   Performance Evaluation

This section evaluates the performance of the win–win cooperative green radio resource allocation mechanism based on the NBS, which is described by the Algorithms 4.2.1–4.2.3, and it is compared with two benchmarks. The first benchmark is the SMS that is given by the radio resource allocation in (4.21), which resembles the radio resource allocation schemes in [32, 33] and [36]. The second benchmark is the NCS that is given by the radio resource allocation in (4.1), which resembles the radio resource allocation strategies in [17, 38, 79, 111, 147–150] and [151]. Simulations are performed for a geographical region that is covered by two BSs belonging to two different networks. The angle between the horizontal axis and the line joining the two BSs is randomly chosen following a uniform distribution in $[0, 2\pi]$. Each BS has a maximum power constraint of 191 W, static power consumption of 77 W, and drain efficiency of 35% for the power amplifier, that is, $1/\xi = 0.35$ [152]. MTs are uniformly distributed in the geographical region with 50 and 30 MTs as the subscribers of the first and second network, respectively. All MTs require the same minimum data rate 0.1 Mbps. Furthermore, MTs are subject to independent Rayleigh fading channels and path-loss exponent $\alpha = 5$ with both BSs. The one-sided noise power spectral density is $N_0 = -174$ dBm/Hz.

We present three sets of results. In the first set, which is depicted in Figures 4.3 and 4.4, it is assumed that all the subscribers of the first network enjoy good channel conditions with both BSs, while the number of second network subscribers who enjoy good channel conditions with the first network BS is varied. As a result, all network 1 subscribers can connect to both BSs, while only a fraction Σ_2 of network 2 subscribers can connect to the BS of the first network. The rest of the second network subscribers who cannot connect to the first network BS receive their required radio resources only from their home network, that is, network 2. In this set of results, the BSs are separated by 250 m. Also, the total bandwidth available at each BS is 10 MHz. Hence, both networks have equal bargain power $\varrho_n = 0.5$ for $n \in \{1, 2\}$. The second set of results, which is given by Figures 4.5 and 4.6, investigates the system performance for different bargain powers of the networks. In particular, we vary the total bandwidth available for network 2 in the range $[5, 50]$ MHz and fix the total bandwidth available for network 1 to 10 MHz. As a result, the bargain power of network 2 varies in the range $\varrho_2 \in [0.5, 0.833]$ and $\varrho_1 = 1 - \varrho_2$. In this set of results, all MTs are able to connect to both networks using their multi-homing capabilities. Again, the distance between both BSs is 250 m. The last set of results, which is depicted in Figures 4.7 and 4.8, assumes a variable distance between the two BSs. Hence, we fix the position of the second BS and move the first BS away from it within a distance range of $[50, 400]$ m. The total bandwidth available at each BS is 10 MHz and all MTs are able to connect to both networks using their multi-homing capabilities. The three sets of results represent situations where cooperation is not always beneficial for network 2. Simulation results present the confidence interval for a 95% confidence level.

Figure 4.3 shows a plot of total power consumption in the geographical region, $\sum_n P_n$ against the number of network 2 subscribers who can connect to the BSs of both networks, Σ_2. The NCS total power consumption is independent of Σ_2 as in the NCS each network allocates its radio resources only to its own subscribers, regardless of the fraction of users who are able to connect to both networks using multi-homing. For the cooperative solutions (SMS and NBS), the total power consumption in the geographical region is reduced as more subscribers from network 2 are able to connect to both BSs. This is because, with Σ_2, more opportunities are created by the cooperative solutions through multi-homing to satisfy the users' required QoS at reduced network power consumption. While both cooperative solutions achieve lower power consumption compared to the NCS, only the NBS offers cooperation incentives for power minimization in comparison with the SMS, as will be shown next.

Figure 4.4 shows a plot of the total power consumption for each network BS, P_{ns} against Σ_2. For the entire range of Σ_2, the SMS reduces the power consumption for network 1 BS (SMS1 < NCS1) significantly at the cost of an increase in the power consumption of network 2 BS (SMS2 > NCS2). For $\Sigma_2 \in [0, 15]$, SMS1 is decreasing with Σ_2, since more power saving opportunities can be achieved with Σ_2 within this range. For $\Sigma_2 \in (15, 30]$, SMS1 slightly increases with Σ_2, as more users from network 1 now can receive part of their required resources from network 1; however, still SMS1 is significantly less than NCS1, yet SMS2 >

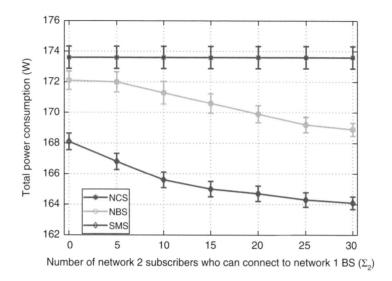

Figure 4.3   Total power consumption in the geographical region with different Σ_2 [137]. The BSs are separated by 250 m. The total bandwidth available at each BS is 10 MHz

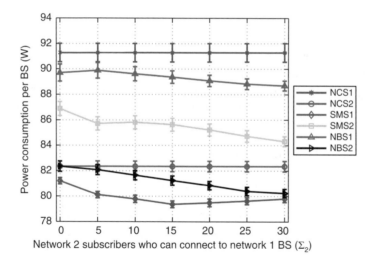

Figure 4.4   Power consumption for each BS with different Σ_2 [137]. The BSs are separated by 250 m. The total bandwidth available at each BS is 10 MHz

NCS2. Consequently, within $\Sigma_2 \in [0, 30]$, the SMS does not offer an incentive for network 2 BS to cooperate with network 1 BS for power minimization. On the contrary , the NBS always results in a power consumption for both BSs (NBS1 for BS1 and NBS2 for BS2) that is less than or equal to the NCS. Consequently, the NBS always guarantees mutual benefit of power saving for both networks. Hence, within the win–win cooperative framework, both networks have incentives to cooperate for power saving.

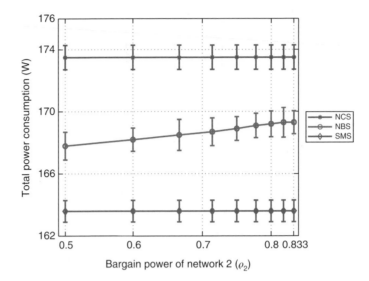

Figure 4.5 Total power consumption in the geographical region with different ϱ_2 [137]. The distance between both BSs is 250 m. The total bandwidth available at BS 1 is 10 MHz and BS 2 is in the range [5, 50] MHz

Figure 4.5 shows the total power consumption in the geographical region for different bargain powers of the networks. Both the NCS and SMS are independent of the bargain power of the network. For the NBS, as the bargain power of network 2 increases, less power saving is achieved in the geographical region. The rationale behind such a behaviour is explained using the next result.

Figure 4.6 shows the total power consumption for each network BS versus ϱ_2. The NBS represents a weighted proportional fair solution. For a larger value of ϱ_2, more emphasis is given to power saving for the second network BS. This is met by lowering the importance of power saving at the first network BS. While such a behaviour limits the power saving opportunities, which is clear in Figure 4.5 for the NBS compared with the SMS, it gives a flexibility to the cooperating entities to achieve a better bargain (power saving) based on their capabilities (e.g. total available bandwidth). This is clear from the behaviour of NBS1 and NBS2 compared with NCS1 and NCS2, respectively, for different bargain power values. On the contrary, for the SMS, regardless of the capabilities of network 2 (total available bandwidth as compared with network 1), it always achieves a higher power consumption than the NCS (as given by SMS2 compared with NCS2), which does not motivate the second network to cooperate, unlike the NBS (as given by NBS2 compared with NCS2).

Figure 4.7 shows the total power consumption in the geographical region for different separation distances among the two BSs. As the first network BS moves further away from the second network BS, the total power consumption increases for both the NCS and NBS, while it first decreases for the SMS and then saturates. Such a behaviour is explained using the following result.

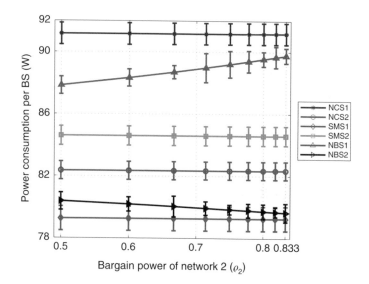

Figure 4.6 Power consumption for each BS with different ϱ_2 [137]. The distance between both BSs is 250 m. The total bandwidth available at BS 1 is 10 MHz and BS 2 is in the range $[5, 50]$ MHz

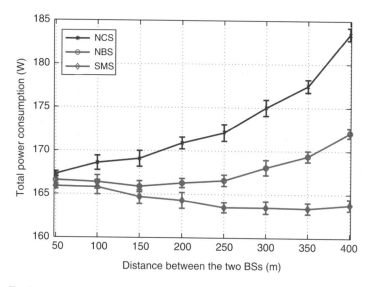

Figure 4.7 Total power consumption in the geographical region with different separation distances between the two BSs [137]. The total bandwidth available at each BS is 10 MHz

Figure 4.8 shows a plot of the total power consumption for each network BS against the separation distance between the two BSs. For the NCS, as the first BS moves further away from the second BS, the distance between some subscribers of the first network and their home network (network 1 BS) increases, leading to an increased transmission power for BS 1 (and hence an increased total power consumption for BS 1, i.e. NCS1). On the contrary, the

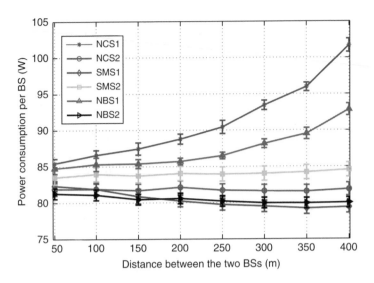

Figure 4.8 Power consumption for each BS with different separation distances between the two BSs [137]. The total bandwidth available at each BS is 10 MHz

transmission power for BS 2 is not much affected due to the fixed location of BS 2. Consequently, the total power consumption in the region increases with the separation distance (as shown in Figure 4.7 for the NCS). For the SMS, as the separation distance increases, the first network BS relies more on the second network BS to serve its faraway subscribers that are in closer proximity to BS 2. Consequently, this increases the transmission power of BS 2 (and hence increases BS 2 total power consumption, i.e. SMS2), while it decreases the transmission power of BS 1 (and hence decreases BS 1 total power consumption, i.e. SMS1). Overall, this decreases the power consumption in the region with the separation distance (as shown in Figure 4.7 for the SMS). While the SMS reduces power consumption of BS 1 (SMS1 < NCS1), it increases the power consumption of BS2 (SMS2 > NCS2). Finally, the NBS always guarantees that the power consumption of both BSs is lower than that of the NCS. In order to keep the second network BS power consumption lower than that corresponding to the NCS (NBS2 < NCS2), the first network BS has to increase its transmission power to cope up with the increased distance between BS 1 and some MTs; however, BS 1 still achieves lower power consumption than the NCS (NBS1 < NCS1). Hence, this increases the power consumption in the region with the separation distance (as shown in Figure 4.7 for the NBS).

4.3 IDC Interference-Aware Green Resource Allocation

In this section, we present a minimum power consumption radio resource allocation algorithm for LTE/WiFi networks that takes into account the mutual IDC interference between LTE and WiFi. Such a resource allocation algorithm fits the scenario, where different radio interfaces of the MT run different applications, and hence, the uplink transmission of one radio interface causes interference on the downlink reception of another radio interface that operates in a close frequency band. This is different from the scenario described in the previous section, where all radio interfaces of a given MT serve the downlink reception of one application, and hence, no IDC interference exists.

Consider a network consisting of a WLAN AP overlaid in the coverage of a single-cell LTE BS, whereby M MTs with LTE and WLAN radio interfaces are distributed in the region of interest. Under the typical scenario, where users run different applications on LTE and WLAN interfaces, we consider a scenario where LTE and WLAN operate in adjacent bands in a TDD mode. Therefore, mutual IDC interference exists between the LTE and WLAN interfaces of each user. The LTE BS has a total bandwidth of B MHz and a set of channels $C = \{1, 2, \cdots, C\}$. The total bandwidth is equally divided among the C channels, each having B/C channel bandwidth. The channel is assigned to a single user at a given time. However, a user can be assigned to more than one channel. We denote $x_{c,m}$ as the channel assignment variable for user m in channel c, where $x_{c,m} = 1$ if channel c is assigned to user m, otherwise, $x_{c,m} = 0$. Variables $P_{c,m}^{\mathrm{DL}}$ and $h_{c,m}$ denote the allocated downlink power and channel gain for user m on channel c, respectively. Let N_0 and $I_{c,m}$ denote the noise power spectral density and IDC interference on LTE channel c assigned to user m, respectively. Using Shannon formula, the maximum achievable data rate for user m in the LTE network downlink can be expressed by

$$R_m = \sum_{c=1}^{C} x_{c,m} \frac{B}{C} \log_2 \left(1 + \frac{P_{c,m}^{\mathrm{DL}} h_{c,m}}{N_0 + I_{c,m}} \right). \tag{4.22}$$

For the WLAN, each AP is assigned a single channel and users share this channel in a TDMA manner. This can be achieved by an enhanced version of the DCF [130]. The user is assigned the whole AP bandwidth (B') in its allocated time fraction denoted by t'_m. Hence, the WLAN downlink data rate of user m is expressed as

$$R'_m = t'_m B' \log_2 \left(1 + \frac{P_m^{\mathrm{DL}'} h'_m}{N'_0 + I'_m} \right), \tag{4.23}$$

where $P_m^{\mathrm{DL}'}$ and h'_m denote the allocated downlink power and channel gain for MT m, respectively, N'_0 is the noise power at the WLAN receiver, and I'_m represents the LTE interference on the WLAN signal.

The IDC interference that affects LTE channel c when it is allocated to user m, $I_{c,m}$, and the IDC interference at the WLAN receiver, I'_m, are given in (3.2) and (3.4), respectively, as function of the interfering WLAN uplink power allocated by MT m, P_m^{UL}, the interfering LTE uplink power allocated by MT m on channel c, $P_{c,m}^{\mathrm{UL}'}$ and the losses due to the antenna isolation at user m, A_m.

4.3.1 IDC Interference-Aware Resource Allocation Design

The WLAN and LTE uplink transmission powers leak into the downlink reception paths of LTE and WLAN, respectively, as depicted in Figure 3.4. In this subsection, we design a resource allocation mechanism that meets the data rate requirements in the victim downlink path by using a joint channel, time and downlink power radio resource allocation ($x_{c,m}$, t'_m, $P_{c,m}^{\mathrm{DL}}$ and $P_m^{\mathrm{DL}'}$). The IDC resource allocation problem is formulated with the target of minimizing the downlink transmission power as follows

$$\min_{\{x_{c,m}, t'_m, P^{DL}_{c,m}, P^{DL'}_m\}} \sum_{m=1}^{M} \sum_{c=1}^{C} P^{DL}_{c,m} + \sum_{m=1}^{M} P^{DL'}_m$$

$$\text{s.t.} \quad R_m \geq R_{m,\min}, \quad \forall m \in \mathcal{M}$$

$$R'_m \geq R'_{m,\min}, \quad \forall m \in \mathcal{M}$$

$$\sum_{m=1}^{M} x_{c,m} \leq 1, \quad \forall c \in \mathcal{C}; \ x_{c,m} \in \{0,1\}$$

$$\sum_{m=1}^{M} t'_m \leq 1, \qquad 0 < t'_m < 1, \ \forall m, \tag{4.24}$$

where $R_{m,\min}$ and $R'_{m,\min}$ are the minimum data rate requirements for the LTE and WLAN radios, respectively. The third constraint in (4.24) ensures that each LTE channel is assigned to a single user. The last constraint in (4.24) guarantees that one WLAN user can access the channel at a given time.

The terms $I_{c,m}$ and I'_m in the data rate expressions (4.22) and (4.23) couple the resource allocation of both radios, since the amount of interference depends on the frequency gap between the LTE and WLAN channels allocated to the user. Under the fact that each transceiver for each network can calculate the interference power in its band using (3.2) and (3.4), (4.24) can be solved by letting each BS/AP perform its resource allocation while considering the IDC observed from the other network.

4.3.1.1   LTE Radio Resource Allocation

Optimizing only for the LTE resources, the decoupled problem has the target of minimizing the LTE transmission power and is constrained by meeting the LTE data rate requirements of all users while considering the IDC interference as follows

$$\min_{\{x_{c,m}, P^{DL}_{c,m}\}} \sum_{m=1}^{M} \sum_{c=1}^{C} P^{DL}_{c,m}$$

$$\text{s.t.} \quad R_m \geq R_{m,\min}, \quad \forall m \in \mathcal{M}$$

$$\sum_{m=1}^{M} x_{c,m} \leq 1, \quad \forall c \in \mathcal{C}; \ x_{c,m} \in \{0,1\}. \tag{4.25}$$

Problem (4.25) is an MINLP due to the existence of the binary optimization variable $x_{c,m}$ and the continuous optimization variable $P^{DL}_{c,m}$. By relaxing the binary constraint $x_{c,m}$ to take continuous values in the range $[0,1]$, the optimum power and channel allocation variables can be expressed as follows [153]

$$P^{DL*}_{c,m^*} = \begin{cases} \left[\dfrac{\lambda_m \frac{B}{C}}{\ln 2} - \dfrac{N_0 + I_{c,m}}{h_{c,m}} \right]^+ & \text{if } x_{c,m_c^*} = 1 \\[4mm] 0 & \text{if } x_{c,m_c^*} = 0, \end{cases} \tag{4.26}$$

and

$$m_c^* = \arg \min_m \aleph_{c,m}(\lambda_m),$$ (4.27)

where

$$\aleph_{c,m}(\lambda_m) =$$

$$\begin{cases} \dfrac{\lambda_m \frac{B}{C}}{\ln(2)} - \dfrac{N_0 + I_{c,m}}{h_{c,m}} - \lambda_m \dfrac{B}{C} \log_2 \left(\dfrac{\lambda_m \frac{B}{C} h_{c,m}}{(N_0 + I_{c,m}) \ln(2)} \right), & \lambda_m > \dfrac{(N_0 + I_{c,m}) \ln(2)}{\frac{B}{C} h_{c,m}} \\ 0 & \text{otherwise.} \end{cases}$$ (4.28)

The optimal values of the Lagrangian multipliers ($\lambda_m > 0$) must satisfy the complementary slackness condition expressed as $\lambda_m^*(R_{m,\min} - R_m^*) = 0$. The resource allocation solution states that the user presenting the minimum value for $\aleph_{c,m}$ is assigned channel c without any time-sharing with other users. This radio resource assignment converges only in low-dense networks, where the number of users is much lower than the number of channels. However, in high-dense networks, where the number of users is comparable to the number of channels, some channels keep oscillating between two or more users and convergence does not occur. In order to overcome this problem, a priority scheme is proposed to settle the competition between users as follows:

- Case 1: Only one user is not assigned any channels among the competing users: The contested channel will be assigned to this user with no other assigned channels.
- Case 2: None of the competing users is assigned a channel: The contested channel will be assigned to the user having the worst channel conditions among the competing users.

The intuition behind this scheme is to allow fairness among users. In Case 1, the winning user is the one with no assigned channels. In Case 2, the channel is assigned to the most needy user who will face even worse channel conditions if it is assigned any channel other than the contested channel.

4.3.1.2   WLAN Radio Resource Allocation

Similarly, in allocating only the WLAN resources, our objective is to minimize the WLAN transmission power while meeting the WLAN data rate requirements of all users and considering the IDC interference. This optimization problem is formulated as follows

$$\min_{\{t_m', P_m^{\text{DL}'}\}} \sum_{m=1}^{M} P_m^{\text{DL}'}$$

$$\text{s.t.} \quad R_m' \geq R_{m,\,\min}', \quad \forall m \in \mathcal{M}$$ (4.29)

$$\sum_{m=1}^{M} t_m' \leq 1, \quad 0 < t_m' < 1, \; \forall m.$$

Problem (4.29) is a convex optimization problem due to the linearity of the objective function and the last constraint together with the concavity of the first constraint. By applying the

Lagrangian duality theory, the optimal power and time fraction allocation can be expressed as follows [127]

$$P_m^{DL'} = \left[\frac{N_0' + I_m'}{h_m'} \left(2^{\frac{\nu'}{\lambda_m' B'}} - 1 \right) \right]^+,$$

$$t_m'^* = \left[\frac{2^{\frac{\nu'}{\lambda_m' B'}} \ln(2)(N_0' + I_m')}{h_m' \lambda_m' B'} \right]^+. \tag{4.30}$$

The optimal values of the Lagrangian multipliers ($\lambda_m' > 0$ and $\nu' > 0$) can be obtained by enforcing the complementary slackness conditions given by $\lambda_m'^*(R_{m,\min}' - R_m'^*) = 0$ and $\nu'^*(\sum_{m=1}^{M} t_m'^* - 1) = 0$.

4.3.2 Performance Evaluation

Consider the simultaneous operation of LTE and WLAN radio technologies in one mobile. The WLAN operates on a 22-MHz TDD channel, while the LTE operates on a 5-MHz TDD channel. The design of the WLAN spectrum mask is obtained from the IEEE 802.11 standard [131]. The used LTE spectrum mask is similar to a 5-MHz ISM filter, but shifted to create a pass-band for the LTE channels as carried out by the 3GPP [128]. Interference parameters are summarized in Table 3.1.

We consider an LTE BS located at coordinates $(0,0)$. One WLAN AP is located within the coverage of the LTE BS at coordinates $(200,0)$ with radius of 100 m. Users with LTE and WLAN interfaces are randomly distributed within the WLAN AP coverage. A Rayleigh fading channel model is used to describe the wireless channel between each user and the BS/AP. For the LTE, we consider 20 channels, each having 5-MHz channel bandwidth between 2,300 and 2,400 MHz. The maximum allowable downlink power per channel is 33 dBm. The AP operates on IEEE 802.11 channel 3 (centre frequency equal to 2,422 MHz). WLAN users share this channel in the time domain. The maximum allowable WLAN downlink power for each user is 27 dBm. The rest of simulation parameters are depicted in Table 4.1. We compare the IDC interference-aware resource allocation mechanism with a benchmark defined in [153], which overlooks the IDC interference.

Table 4.1 Simulation parameters [127]

Parameter	Value
LTE path-loss model	$35.3 + 37.6\log(d)$
WLAN path-loss model	$38.2 + 30\log(d)$
LTE noise figure	9 dBm
LTE noise power density	-174 dBm/Hz
WLAN noise power	-90 dBm
Rate requirement per LTE user ($R_{m,\min}$)	5 Mbps
Rate requirement per WLAN user ($R_{m,\min}'$)	2 Mbps

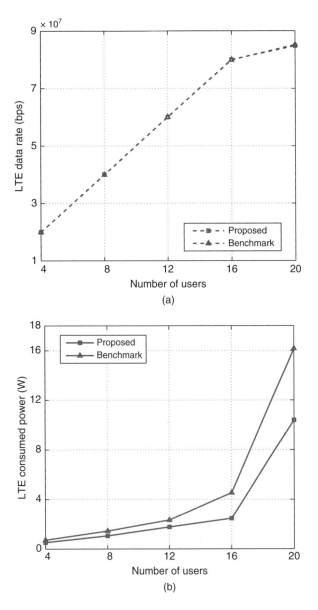

Figure 4.9   LTE network performance [127]. (a) Achieved data rate; (b) power consumption

Figure 4.9 depicts the achieved data rate and consumed power in the LTE network. As expected, the IDC interference-aware mechanism outperforms the benchmark in terms of power consumption. However, both mechanisms achieve the same data rate. In the IDC interference-aware mechanism, the users are assigned channels having the least interference effects and fading conditions. In the benchmark, the user selects the channels based on the

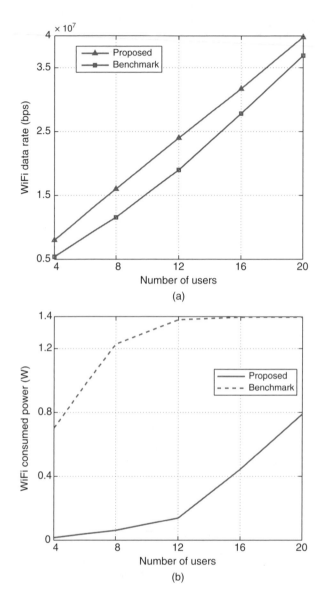

Figure 4.10 WLAN performance [127]. (a) Achieved data rate; (b) power consumption

fading conditions only without taking into consideration that some of these channels may experience high IDC interference. When the user is assigned a highly interfered channel, it keeps increasing its transmission power without achieving the required data rate till it is assigned additional channels based on the water-filling principle. Then, using these additional channels, the user meets the required throughput at the expense of wasting some of its power on the highly interfered channels.

The IDC interference-aware LTE resource allocation mechanism uses the least interfered LTE channels, which present sufficient frequency separation from the WLAN channel. Therefore, the WLAN radio does not experience high interference and achieves high data rate with low power consumption, as depicted in Figure 4.10. By contrast, the LTE benchmark uses the highly interfered channels, which are closest to the WLAN band. These channels produce high interference to the WLAN channel. Therefore, in the benchmark WLAN resource allocation mechanism, some users consume their maximum transmission power and do not meet the required rate, as depicted in Figure 4.10. The WLAN AP has only one channel to offer to users. When this channel is subjected to high interference, the WLAN performance is severely affected unlike the LTE BS that presents a variety of channels.

4.4 Summary

This chapter presents two radio resource allocation mechanisms for green downlink communications. The first mechanism supports green communications in a multi-operator heterogeneous wireless medium. It guarantees mutual benefits in power saving among different service providers. The mechanism is based on a Nash bargain game among different service providers and is implemented in a decentralized manner. The Nash bargain strategy guarantees mutual benefits among different service providers as compared with the total (sum) power minimization approach, and hence, it offers incentives for different networks to cooperate. The second mechanism accounts for the filtering characteristics of the transceivers as well as their physical characteristics (e.g. antenna isolation) to alleviate the IDC interference, which can severely limit the performance of both the WLAN and LTE networks. The resource allocation mechanism minimizes not only the consumed power but also the effects of IDC interference implicitly.

5

Multi-homing for a Green Uplink

In this chapter, green uplink communications are investigated for battery-constrained MTs with service quality requirements and multi-homing capabilities. A heterogeneous wireless medium is considered, where MTs communicate with BSs/APs of different networks with overlapped coverage. Two application scenarios are studied, namely data uploading and video streaming. First, we give an introduction that summarizes the challenging issues that should be considered while designing a green uplink multi-homing radio resource allocation mechanism for both data uploading and video streaming. Then, two radio resource allocation mechanisms are presented to address the associated challenging issues with both application scenarios.

5.1 Introduction

The past decade has witnessed significant advances in the design of MTs and the offered communication services for mobile users. In particular, MTs are currently equipped with processing and display capabilities that enable them to support voice, video and data calls. In addition, MTs are capable of establishing simultaneous communications with BSs and APs of different networks, through multiple radio interfaces and using the multi-homing feature [154]. Utilizing multiple radio interfaces of an MT to support video or data transmission through multi-homing service can improve service quality in many aspects [155, 156]. Transmitting (data or video) packets over multiple networks (i) increases the amount of aggregate bandwidth available for the application, (ii) reduces the correlation between consecutive packet losses due to transmission errors or network congestion and (iii) allows for mobility support as it can reduce the probability of an outage when a communication link is lost with the current serving network as the user moves out of its coverage area [25].

Such an advancement in wireless services results, however, in high energy consumption for the MTs. It has been shown that there exists an exponential increase in the gap between the MT demand for energy and the offered battery capacity [24]. The operational time of an MT in between battery chargings is considered to be a significant factor in the user-perceived QoS [25, 157, 158]. Besides developing new battery technology with improved capacity, the

Green Heterogeneous Wireless Networks, First Edition. Muhammad Ismail, Muhammad Zeeshan Shakir, Khalid A. Qaraqe and Erchin Serpedin.
© 2016 John Wiley & Sons, Ltd. Published 2016 by John Wiley & Sons, Ltd.

operational period of an MT between battery chargings can be extended by managing its energy consumption [159]. Consequently, energy-efficient (*green*) communication techniques have been proposed as a promising solution to regulate the MT energy usage while satisfying the user-required QoS. In this context, the unique characteristics of the target application should be accounted for while designing such green communication techniques.

In general, the key difference between video and data calls is the impact of the allocated resources on the call presence in the system [160]. For video calls, the amount of the allocated resources influences the perceived video quality experienced on the video terminal, while it does not affect the video call duration. On the contrary, the resource allocation to a data call affects its throughput and thus its duration. Consequently, for data calls, the objective is to maximize the achieved throughput at a reduced energy consumption for the MT. This is equivalent to maximizing the resulting energy efficiency as expressed by (1.28)–(1.32) while satisfying a target data rate for the MT. On the contrary, for video calls, maximizing throughput does not necessarily improve the resulting video quality. If video packets are not properly scheduled for transmission, they might miss their playback deadline, which consequently degrades the achieved video quality. As a result, for video calls, the main objective is to schedule video packets in a way that maximizes the resulting video quality while accounting for the MT battery energy limitation.

In this chapter, we present two radio resource allocation mechanisms that support green multi-homing data and video calls [54, 161]. The next section deals with developing green multi-homing radio resource allocation for data calls based on fractional programming [54] while Section 5.3 investigates developing a green multi-homing resource allocation mechanism for video calls based on a statistical quality guarantee [161].

5.2   Green Multi-homing Uplink Resource Allocation for Data Calls

The main limitation of the existing uplink multi-homing green radio resource allocation mechanisms for data calls in a heterogeneous wireless medium is that these solutions focus only on optimal power allocation to different radio interfaces of the MT, given an allocated bandwidth. Hence, the main focus so far is on exploiting the diversity in fading channels and propagation losses between the MT and different BSs/APs to enhance the uplink energy efficiency. However, further improvement can be achieved by exploiting the disparity in available radio resources at the BSs/APs of different networks. This calls for a joint optimization framework for bandwidth and power allocation to maximize uplink energy efficiency for a set of MTs with multi-homing capabilities. Furthermore, the existing resource aggregation schemes (e.g. carrier aggregation in LTE-advanced [134, 162, 163]) assume the scenario where all resources belong to the same service provider. Hence, centralized resource allocation schemes can be adopted. On the contrary, in a heterogeneous networking environment, the aggregated resources are operated by different service providers. Hence, novel decentralized mechanisms should be investigated to enable coordination among MTs and BSs/APs of different networks to satisfy the target QoS in an energy-efficient manner.

In this section, we present a QoS-based optimization framework for joint uplink bandwidth and power allocation to maximize energy efficiency for MTs in a heterogeneous wireless medium [54]. The heterogeneity of the wireless medium is captured in the problem formulation, in terms of different service areas, channel conditions, available radio resources at BSs/APs of different networks and different maximum transmit powers at the MTs. The

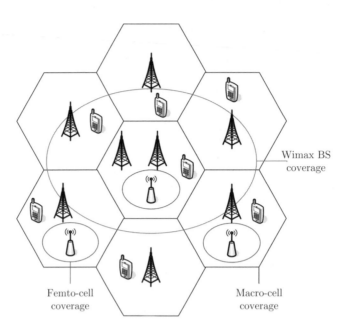

Figure 5.1   Network coverage areas [54]

energy-efficient uplink communication problem is formulated to jointly allocate uplink transmission bandwidth and power to a set of MTs, with minimum required QoS and multi-homing capabilities, from a set of BSs/APs with overlapped coverage. In dealing with a multi-user system, the objective is to maximize the performance of an MT that achieves the minimum energy efficiency.

A geographical region is considered where a set $\mathcal{N} = \{1, 2, \ldots, N\}$ of wireless networks is available, as shown in Figure 5.1. Different networks are operated in separate frequency bands by different service providers, and consequently, no interference exists among these networks. In particular, the set $\mathcal{N}$ contains cellular networks with heterogeneous cell sizes (e.g. macro, pico and femto cells) and overlapped coverage areas. Each network $n \in \mathcal{N}$ has a set $\mathcal{S}_n = \{1, 2, \ldots, S_n\}$ of BSs/APs in the geographical region. Interference management schemes (e.g. soft frequency reuse [138–141]) are implemented for interference mitigation among BSs/APs within the same network. Due to the overlapped coverage from BSs/APs of different networks, the geographical region is partitioned into a set of service areas. A unique subset of BSs/APs covers each service area. The total bandwidth available at network n BS/AP s is denoted by B_{ns}. A cooperative networking scenario is considered where different networks in $\mathcal{N}$ cooperate in radio resource allocation through signalling exchange over a backbone [1].

A set of MTs $\mathcal{M} = \{1, 2, \ldots, M\}$ perform uplink multi-homing data transmission in the geographical region. Let $\mathcal{M}_{ns} \subseteq \mathcal{M}$ denote the subset of MTs which lie in the coverage area of network n BS/AP s. Using the multiple radio interfaces and through the multi-homing capability, each MT can communicate with multiple BSs/APs simultaneously. The bandwidth allocated on the uplink from network n BS/AP s to MT m is denoted by B_{nsm}, where

$B_{nsm} = 0$ for $m \notin \mathcal{M}_{ns}$.[1] Let P_{nsm} represent the transmission power allocated by MT m to its radio interface that communicates with network n BS/AP s. Denote ξ by the power amplifier efficiency. Hence, the MT transmission power consumption at each radio interface is given by P_{nsm}/ξ [16]. The MT circuit power consumption for each radio interface consists of two components. The first component P_{nsm}^{F} is a fixed circuit power consumption for each MT radio interface, and it captures the power consumption of the RF chain, that is, digital-to-analog converter, RF filter, local oscillator and mixer. The second component is a dynamic part that refers to the digital circuit power consumption and scales with the allocated transmission bandwidth (as bandwidth increases, more computations and baseband processing are required). The dynamic component is expressed as [53]

$$P_{nsm}^{D} = P_{D}^{ref} + \sigma_{nsm} \frac{B_{nsm}}{B_{ref}}, \qquad (5.1)$$

where P_{D}^{ref} denotes the reference digital circuit power consumption for a reference bandwidth B_{ref} and σ_{nsm} is a proportionality constant. For $m \notin \mathcal{M}_{ns}$, $P_{nsm} = P_{nsm}^{F} = P_{nsm}^{D} = 0$. Denote $P_{nsm}^{F} + P_{D}^{ref}$ by P_{nsm}^{C} and σ_{nsm}/B_{ref} by ψ_{nsm}. Hence, the MT total power consumption for each radio interface is given by

$$P_{nsm}^{T} = \frac{P_{nsm}}{\xi_{nsm}} + P_{nsm}^{C} + \psi_{nsm} B_{nsm}. \qquad (5.2)$$

Due to technology limitations, each MT radio interface has a maximum transmission power P_{ns}^{T}. The maximum power constraint at MT m is given by P_{m}^{T}. The MT target service quality can be obtained using the minimum data rate R_{m}^{min} for MT m.

The channel power gain between MT m and network n BS/AP s is denoted by h_{nsm}, and it captures both the wireless channel Rayleigh fading and path loss. Let d_{nsm} denote the distance between MT m and network n BS/AP s. The associated path loss is given by $d_{nsm}^{-\alpha}$, where α is the path-loss exponent. Let κ_{nsm} be a Rayleigh random variable associated with the link between MT m and network n BS/AP s. The channel power gain between MT m and network n BS/AP s is given by

$$h_{nsm} = \kappa_{nsm} d_{nsm}^{-\alpha}. \qquad (5.3)$$

The one-sided noise power spectral density is denoted by N_0.

5.2.1 Optimal Green Uplink Radio Resource Allocation with QoS Guarantee

According to Shannon formula, the data rate achieved by MT m using the radio interface communicating with network n BS/AP s is given by

$$R_{nsm} = B_{nsm} \log_2 \left(1 + \frac{P_{nsm} h_{nsm}}{N_0 B_{nsm}} \right), \qquad \forall n, s, m. \qquad (5.4)$$

---

[1] Since in this chapter we focus only on an uplink scenario, we omit the superscripts UL and DL from all mathematical symbols, unlike our definitions in Chapter 3.

The total achieved data rate by MT m is $R_m = \sum_n \sum_s R_{nsm}$, which should satisfy the required QoS, that is,

$$R_m \geq R_m^{\min}, \quad \forall m. \tag{5.5}$$

The total allocated bandwidth by network n BS/AP s should not be larger than the total available bandwidth, that is,

$$\sum_{m \in \mathcal{M}_{ns}} B_{nsm} \leq B_{ns}, \quad \forall n, s. \tag{5.6}$$

Given the technical limitation on the maximum transmission power for each radio interface, we have

$$P_{nsm} \leq P_{ns}^{\mathrm{T}}, \quad \forall n, s, m. \tag{5.7}$$

The MT total power consumption includes both data transmission and circuit power consumption for all active radio interfaces, that is, for MT m, $P_m = \sum_n \sum_s P_{nsm}^{\mathrm{T}}$. The total power consumption for MT m, P_m, should satisfy the MT maximum power constraint, that is,

$$P_m \leq P_m^{\mathrm{T}}, \quad \forall m \in \mathcal{M}. \tag{5.8}$$

Define the energy efficiency of MT m, η_m, as the ratio of the total achieved data rate to the total power consumption, that is, $\eta_m = R_m/P_m$. The objective is to maximize the minimum achieved energy efficiency η_m for $m \in \mathcal{M}$. This is obtained through joint bandwidth and power allocation from all networks in $\mathcal{N}$ to all MTs in $\mathcal{M}$, while satisfying the required minimum transmission rates and the total bandwidth and power constraints. Hence, the problem is formulated as

$$\max_{B_{nsm}, P_{nsm}} \quad \left\{ \min_{m \in \mathcal{M}} \eta_m \right\}$$
$$\text{s.t.} \quad (5.5)\text{--}(5.8), \tag{5.9}$$
$$B_{nsm}, P_{nsm} \geq 0, \quad \forall n, s, m.$$

Problem (5.9) is classified as a *max–min fractional program* [164]. The optimization problem (5.9) is a concave–convex fractional program, since the numerator of η_m, that is, R_m, is concave, the denominator is convex and the constraints constitute a convex set in B_{nsm} and P_{nsm} [54]. In order to solve (5.9), the following steps can be applied [54].

5.2.1.1 Transform (5.9) into a Convex Optimization Problem

This can be done following the parametric approach using a given parameter λ [164]. The optimal value of λ, which results in the optimal bandwidth and power allocation for (5.9), can be obtained through an iterative algorithm. For a non-negative parameter $\lambda = \min_{m \in \mathcal{M}} \eta_m$, (5.9) can be transformed into

$$F(\lambda) = \max_{B_{nsm}, P_{nsm}} \quad \left\{ \min_{m \in \mathcal{M}} \{ R_m - \lambda P_m \} \right\}$$
$$\text{s.t.} \quad (5.5)\text{--}(5.8), \tag{5.10}$$
$$B_{nsm}, P_{nsm} \geq 0, \quad \forall n, s, m.$$

The optimal solution of (5.9) can be determined by finding a root of equation $F(\lambda) = 0$, which can be obtained using a Dinkelbach-type algorithm, as given in Algorithm 5.2.4 [165].

Algorithm 5.2.4 Dinkelbach-Type Procedure

Initialization: $\{B_{nsm}(1), P_{nsm}(1)\} > 0 \ \forall n, s, m, \ \lambda(1) = \min\limits_{m \in \mathcal{M}} \eta_m, \ i = 1;$

while $F(\lambda(i)) \neq 0$ **do**

    Solve (5.10) for optimal $\{B_{nsm}(i), P_{nsm}(i)\}$;

    $\lambda(i+1) = \min\limits_{m \in \mathcal{M}} \eta_m(i);$

    $i \longleftarrow i + 1;$

end while

Output:$B_{nsm}, P_{nsm} \ \forall n, s, m.$

Algorithm 5.2.4 converges to the optimal solution of (5.9) in a finite number of iterations [165].

5.2.1.2 Finding the Optimal QoS-Aware Joint Power and Bandwidth Allocation for a Given λ

This is done by solving (5.10) which constitutes an important step in Algorithm 5.2.4. Letting $\theta = \min\limits_{m \in \mathcal{M}} \{R_m - \lambda P_m\}$, (5.10) can be re-written as

$$
\begin{aligned}
\max_{B_{nsm}, P_{nsm}} \quad & \theta \\
\text{s.t.} \quad & \theta \leq R_m - \lambda P_m, \quad \forall m \\
& (5.5)\text{--}(5.8), \\
& B_{nsm}, P_{nsm} \geq 0, \quad \forall n, s, m.
\end{aligned}
\tag{5.11}
$$

Since (5.11) has a linear objective function and convex constraints, it is a convex optimization problem [142]. Following the decomposition theory and using the same steps as those adopted in Chapter 4, the Lagrangian function for (5.11) can be defined as $L(B, P, \mu, \phi, \beta, \omega, \nu)$, where B and P are resource allocation matrices with $B = \{B_{nsm} | \forall n, s, m\}$ and $P = \{P_{nsm} | \forall n, s, m\}$ and $\mu = \{\mu_m | \forall m\}$ is a Lagrangian multiplier vector for the first constraint in (5.11), $\phi = \{\phi_m | \forall m\}$, $\beta = \{\beta_{ns} | \forall n, s\}$, $\omega = \{\omega_{nsm} | \forall n, s, m\}$ and $\nu = \{\nu_m | \forall m\}$ are Lagrangian multiplier vectors and matrices for constraints (5.5)–(5.8), respectively. Applying the KKT conditions on the Lagrangian function and using the same steps as those in Chapter 4, we find (i) the optimal power allocation at the MT as a function of the Lagrangian multipliers ϕ, μ, ω and ν and (ii) the optimal bandwidth allocation at each BS/AP as a function of the Lagrangian multipliers ϕ, μ, ν and β [54]. The optimal values of the Lagrangian multipliers can be obtained by solving the dual problem using a gradient descent method as described in Chapter 4 [54]. In the following, $\delta_{1,2,3,4}$ are sufficiently small step sizes, $\tilde{\phi}_m = \phi_m \sum_m \mu_m$, $\tilde{\nu}_m = \nu_m \sum_m \mu_m$, $\tilde{\omega}_{nsm} = \omega_{nsm} \sum_m \mu_m$ and $\tilde{\beta}_{ns} = \beta_{ns} \sum_m \mu_m$. Algorithm 5.2.5 finds the optimal power allocation for a given allocated bandwidth and λ.

Algorithm 5.2.5 Power Allocation to Each Radio Interface for Every MT m

Input: $B_{nsm} \ \forall n, s, \mu_m, \phi_m$, and λ;

Initialization: $\omega_{nsm}(1) \geq 0$ and $\nu_m(1) \geq 0, i = 1, J = 1$;

while $J = 1$ **do**

  for $n \in \mathcal{N}$ **do**

    for $s \in \mathcal{S}_n$ **do**

$$P_{nsm}^* = B_{nsm} \left[\frac{\mu_m + \tilde{\phi}_m}{\ln(2)\{\frac{1}{\xi}(\lambda\mu_m + \tilde{\nu}_m) + \tilde{\omega}_{nsm}^*\}} - \frac{N_0}{h_{nsm}} \right]^+;$$

$$\omega_{nsm}(i+1) = [\omega_{nsm}(i) - \delta_1(P_{ns}^T - P_{nsm}(i))]^+;$$

    end for

  end for

$$\nu_m(i+1) = \left[\nu_m(i) - \delta_2 \left(P_m^T - \sum_{n \in \mathcal{N}} \sum_{s \in \mathcal{S}_n} \left\{\frac{P_{nsm}(i)}{\xi} + P_{nsm}^C + \psi_{nsm}B_{nsm}\right\}\right)\right]^+;$$

if $|P_{nsm}(i) - P_{nsm}(i-1)| \leq \epsilon$ **then**

  $J = 0$;

else

  $i \longleftarrow i + 1$;

end if

end while

Output: $P_{nsm}^* \ \forall n, s$.

Given the allocated power from Algorithm 5.2.5, the optimal bandwidth can be allocated using Algorithm 5.2.6.

Algorithm 5.2.6 Bandwidth Allocation at Each Network BS/AP to Each MT

Input: $P_{nsm}^* \ \forall n, s, m, \nu_m^*, \phi_m$, and μ_m;

Initialization: $\beta_{ns}(1) \geq 0, i = 1, J = 1$;

while $J = 1$ **do**

  Find $B_{nsm}^*(i)$ as the positive real root of $\log_2\left(1 + \frac{P_{nsm}^* h_{nsm}}{N_0 B_{nsm}^*}\right) -$

$$\frac{P_{nsm}^* h_{nsm}}{\ln(2)(N_0 B_{nsm}^* + P_{nsm}^* h_{nsm})} = \frac{\psi_{nsm}(\lambda\mu_m + \tilde{\nu}_m) + \tilde{\beta}_{ns}}{\mu_m + \tilde{\phi}_m};$$

$$\beta_{ns}(i+1) = [\beta_{ns}(i) - \delta_3(B_{ns} - \sum_{m \in \mathcal{M}_{ns}} B_{nsm}(i))]^+;$$

if $|B_{nsm}(i) - B_{nsm}(i-1)| \leq \epsilon$ **then**

  $J = 0$;

else

  $i \longleftarrow i + 1$;

end if

end while

Output: $B_{nsm}^* \ \forall n, s$.

Using the optimal power and bandwidth allocated in Algorithms 5.2.5 and 5.2.6, the objective now is to find the jointly allocated resources that satisfy the target data rate and maximize the resulting energy efficiency for a given λ. Define $f_m = R_m - \lambda P_m$. Algorithm 5.2.7 gives the optimal solution of (5.11) for a given value of λ by iterating over P_{nsm} and B_{nsm} until convergence to find the optimal joint bandwidth and power allocation solution that maximizes the minimum energy efficiency for a given λ in the region and satisfies the required QoS by all MTs.

Algorithm 5.2.7 Joint Bandwidth and Power Allocation for a Given λ

Input: λ;

Initialization: $\phi_m \geq 0$ and $\mu_m \geq 0 \; \forall m$, P_{nsm} and $B_{nsm} \; \forall n, s, m$, $i = 1$, $K = 1$;

while $K = 1$ **do**

    Every MT broadcasts to all serving BSs/APs its $\phi_m(i)$ value;

    Initialization: $J = 1$;

    while $J = 1$ **do**

        Every MT broadcasts to all serving BSs/APs its η_m, $f_m(i)$, and $P_{nsm}(i)$ values;

        Every BS/AP determines $\theta_{ns}(i) = \min\limits_{m \in \mathcal{M}_{ns}} f_m(i)$;

        All BSs/APs exchange information regarding $\theta_{ns}(i)$ and determine $\theta(i) = \min\limits_{n,s} \theta_{ns}(i)$;

        if for every m, $f_m(i) = \theta(i)$ or $f_m(i) > \theta(i)$ for m with $\mu_m(i) = 0$ **then**

            $J = 0$;

        else

            $\mu_m(i+1) = [\mu_m(i) - \delta_4(R_m(i) - \lambda P_m(i) - \theta)]^+, \forall m$;

            All BSs/APs exchange their information to find $\sum_m \mu_m(i)$ and broadcast this value to all MTs;

            All BSs/APs allocate bandwidth to all MTs using Algorithm 5.2.6;

            All MTs allocate power to their radio interfaces using Algorithm 5.2.5;

        end if

    end while

    if For every m, $|R_m(i) - R_m(i-1)| \leq \epsilon$ **then**

        $K = 0$;

    else

        $\phi_m(i+1) = [\phi_m(i) - \delta_5(R_m(i) - R_m^{\min})]^+, \forall m$;

        $i \longleftarrow i + 1$;

    end if

end while

Output: B_{nsm}^*, P_{nsm}^*, $\forall n, s, m$.

The framework described in Algorithms 5.2.4–5.2.7 is illustrated in Figure 5.2. Using Algorithms 5.2.4–5.2.7, the decentralized uplink energy-efficient radio resource allocation framework is summarized in the following 10 steps:

Step 1. The BSs/APs start with initial bandwidth allocation to all MTs in service and initialize the μ_m value for every MT. The MTs allocate initial powers to their different radio interfaces. Every MT calculates its initial η_m value and broadcasts it together with an initial ϕ_m value to the serving BSs/APs.

Step 2. The BSs/APs exchange their information to find the value $\lambda = \min\limits_{m \in \mathcal{M}} \eta_m$, as shown in Algorithm 5.2.4.

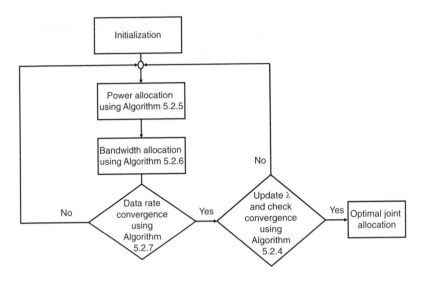

Figure 5.2   Illustration of the framework described in Algorithms 5.2.4–5.2.7

Step 3. The BSs/APs check a termination condition (as shown in Algorithm 5.2.4): $F(\lambda) = 0$?
If the condition is true, the framework is terminated; otherwise, go to Step 4.

Step 4. Every MT broadcasts to its serving BSs/APs its f_m and P_{nsm} values. All BSs/APs
exchange their information regarding their minimum f_m and determine $\theta = \min_m f_m$, as
shown in Algorithm 5.2.7.

Step 5. The BSs/APs check a termination condition (as shown in Algorithm 5.2.7): $f_m = \theta \,\forall m$
with $\mu_m > 0$ and $f_m > \theta \,\forall m$ with $\mu_m = 0$? If the condition is true, go to Step 9; otherwise,
all BSs/APs update the μ_m values, as shown in Algorithm 5.2.7. Also, all BSs/APs exchange
their information to find $\sum_m \mu_m$ and broadcast this value to all MTs.

Step 6. All BSs/APs allocate their radio resources (e.g. bandwidth) to all MTs in service using
Algorithm 5.2.6.

Step 7. All MTs perform power allocation at their radio interfaces using Algorithm 5.2.5.

Step 8. Go to Step 4.

Step 9. Every MT checks its total achieved data rate R_m. If R_m does not converge, the MTs
update their ϕ_m value and broadcast it to all serving BSs/APs, as shown in Algorithm 5.2.7.
Go to Step 4.

Step 10. If R_m converges, every MT transmits its η_m value to all serving BSs/APs. Go to
Step 2.

5.2.2   Suboptimal Uplink Energy-Efficient Radio Resource Allocation

Let I_D denote the number of iterations required for the convergence of the Dinkelbach-type
procedure given in Algorithm 5.2.4. The computational complexity of the optimal radio
resource allocation algorithm is given by $O(I_D M^2 \sum_n S_n)$ [54]. Consequently, the optimal
framework has an online computational complexity that is quadratic in the number of MTs
M. In a system with a large M, the online computational complexity will be high, which

could make it infeasible for the algorithm to run every time the channel state information (CSI) is updated.

In order to further reduce the associated signalling overhead and computational complexity, in the following, we present a suboptimal framework [54], where every time the CSI changes, the radio resource allocation has to be updated. This incurs high signalling overhead over both the backbone connecting the BSs/APs and the air interfaces. In order to reduce the associated signalling overhead and computational complexity, a two-step suboptimal framework is presented.

Step 1: Initialization Phase: In a Rayleigh fading channel, the mean of η_m is given by Ismail et al. [54]

$$\mathbb{E}\{\eta_m\} = \frac{\sum_n \sum_s \frac{B_{nsm}}{2} \log_2(1 + \frac{2\Omega_{nsm} P_{nsm}}{N_0 B_{nsm}})}{\sum_n \sum_s \{\frac{P_{nsm}}{\xi_{nsm}} + P_{nsm}^C + \psi_{nsm} B_{nsm}\}}, \tag{5.12}$$

where $\Omega_{nsm} = \mathbb{E}\{h_{nsm}\}$ and $\mathbb{E}\{\cdot\}$ denotes the expectation. The average QoS constraint is given by

$$\sum_{n \in \mathcal{N}} \sum_{s \in \mathcal{S}_n} \frac{B_{nsm}}{2} \log_2\left(1 + \frac{2\Omega_{nsm} P_{nsm}}{N_0 B_{nsm}}\right) \geq R_m^{\min}. \tag{5.13}$$

Hence, we aim to solve the following optimization problem to set the values of the variables λ, ϕ_m and μ_m [54]

$$\begin{aligned}
\max_{B_{nsm}, P_{nsm}} \quad & \{\min_{m \in \mathcal{M}} \mathbb{E}\{\eta_m\}\} \\
\text{s.t.} \quad & (5.6)-(5.8), (5.13), \\
& B_{nsm}, P_{nsm} \geq 0, \quad \forall n, s, m.
\end{aligned} \tag{5.14}$$

The optimization (5.14) can be solved in a way similar to (5.9). Denote the resulting Lagrangian multipliers from (5.14) by $\bar{\lambda}$, $\bar{\phi}_m$ and $\bar{\mu}_m$.

Step 2: Resource Allocation Update Phase: This phase takes place when the CSI changes. In this step, the power and bandwidth allocations are updated given the instantaneous channel gain h_{nsm} using Algorithms 5.2.5 and 5.2.6, while replacing λ, ϕ_m and μ_m by $\bar{\lambda}$, $\bar{\phi}_m$ and $\bar{\mu}_m$, respectively, as calculated in the initialization phase.

As compared with the optimal framework, the suboptimal framework has a reduced computational complexity. Only Algorithm 5.2.6 is executed at each BS/AP and Algorithm 5.2.5 is executed at each MT for resource allocation update. Almost no signalling exchange takes place during the resource allocation updates, except for the allocated B_{nsm} values that are provided to each MT and the CSI that is updated once during each time slot. While the initialization phase of the suboptimal framework incurs the same computational complexity $O(I_D M^2 \sum_n S_n)$, this is only executed once during the call set-up. The resource allocation update phase that takes place during every time slot has a computational complexity of $O(M \sum_n S_n)$, which is different from the optimal framework. Hence, the resource allocation update, which is executed within every time slot of fixed CSI has an online computational complexity that is linear in M, and it is a more feasible task.

5.2.3 Performance Evaluation

Consider a geographical region that is covered by a micro BS (indexed as 1) and two femto-cell APs (indexed as 2 and 3, respectively). Due to the overlapped coverage among the BS and the two APs, three service areas can be distinguished. In the first and second areas, MTs can get service from both the micro BS and one femto AP. In the third service area, MTs can get service only from the micro BS. The simulation parameters are given in Table 5.1, and are adopted from [16, 45, 53, 166] and [167]. The performance of the optimal and suboptimal approaches is compared with a benchmark based on [31] that investigates only power allocation in a heterogeneous wireless medium for energy efficiency. Hence, given some bandwidth allocation from different networks, every MT independently allocates transmission power to its radio interfaces to maximize its own energy efficiency [54].

Two simulation cases are considered. In the first case, each service area has 5 MTs, and we show the performance of the optimal and suboptimal approaches (using Algorithms 5.2.4–5.2.6 and the two phases in Section 5.2.2, respectively) as compared with the benchmark. In the second case, each service area has 10 MTs. In this case, we show the results of the suboptimal framework as compared with the benchmark. In each of the conducted simulations, we vary the total power consumption at MTs, $P_m^T = [0.5, 3]$, which is displayed across the x-axis. The total available power is used in both data transmission and circuit power consumption. Over the range $P_m^T = [0.5, 3]$, we aim to investigate the performance of the proposed optimal and suboptimal approaches with respect to the benchmark in two situations. The first situation ($P_m^T = [0.5, 1.5]$) presents comparable transmission and circuit power consumption values (due to the low total available power). The second situation ($P_m^T = (1.5, 3]$) presents a large available transmission power than the circuit power consumption (due to the high total available power). Simulation results are averaged over 100 runs.

Figure 5.3a and b show the plots of minimum and average achieved energy efficiencies against P_m^T, respectively. Given the simulation settings, energy efficiency is improved with P_m^T, as the MTs can enhance the achieved throughput at a slight increase in power consumption. With low total available power, lower energy efficiency is achieved due to the comparable values of transmission power consumption (which translates into a useful term, i.e. throughput) and circuit power consumption (which does not contribute to the achieved throughput).

Table 5.1 Simulation parameters [54]

Parameter	Value
B_1	10 MHz
$B_{2,3}$	5 MHz
N_0	-174 dBm/Hz
P_{ns}^T	501.2 mW
P_{nsm}^C	100 mW
R_m^{min}	Uniformly distributed in $[0, 50]$ Mbps
α	4
ξ	0.35
ψ	20×10^{-9} W/Hz

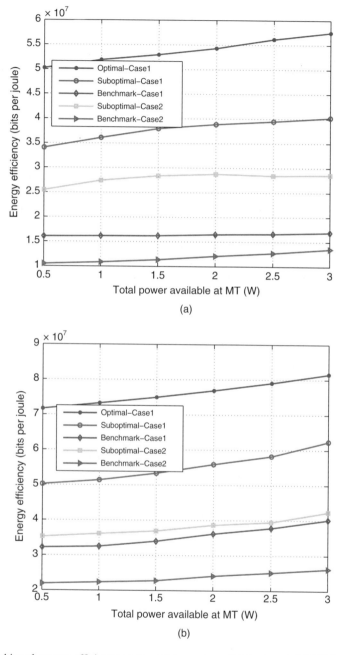

Figure 5.3 Achieved energy efficiency versus total power available at each MT [54]. (a) Minimum achieved energy efficiency; (b) average achieved energy efficiency

With more total available power, more power can be consumed for data transmission, which translates into a higher throughput and enhanced efficiency. As shown in the figures, the proposed optimal and suboptimal approaches outperform the benchmark. This is mainly due to two reasons. First, the proposed approaches jointly optimize bandwidth among MTs and power allocation at each MT to maximize energy efficiency unlike the benchmark, which optimizes only power allocation. Hence, in the new approaches, bandwidth and power allocations are performed according to the channel conditions at different radio interfaces of different MTs and the available energy at each MT. This results in the improved performance in Figure 5.3a and b. Second, the proposed approaches aim to maximize the minimum energy efficiency in the geographical region, unlike the benchmark where every MT aims to maximize its own energy efficiency independent of other MTs. This results in the improved performance of the proposed approaches in Figure 5.3. The optimal approach exhibits improved performance over the sub-optimal approach due to the fact that the optimal approach calculates its dual variables at every time slot using the actual CSI, whereas the suboptimal approach is based on the average CSI. However, overall the suboptimal approach has an improved performance over the benchmark with a reduced signalling overhead and computational complexity. Furthermore, as the number of MTs increases in the system, lower energy efficiency is achieved. This is mainly due to the increased competition over the radio resources at the BS and APs, which leads to reduced bandwidth allocation per user, and hence, a lower energy efficiency is achieved.

Figure 5.4 shows the average satisfaction index of MTs versus P_m^{T}. The satisfaction index captures the ability of the radio resource allocation approaches to satisfy the QoS requirements of the MTs. In particular, the satisfaction index is defined as [83]

$$\mathrm{SI} = \mathbb{E}\left\{ \mathbf{1}_{\mathrm{R}_m \geq \mathrm{R}_m^{\min}} + \mathbf{1}_{\mathrm{R}_m < \mathrm{R}_m^{\min}} \frac{\mathrm{R}_m}{\mathrm{R}_m^{\min}} \right\}, \tag{5.15}$$

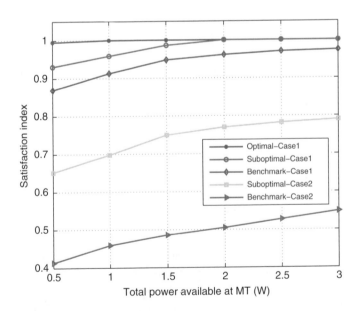

Figure 5.4 Average achieved satisfaction index versus total power available at any MT [54].

where $1_a = 1$ if a is satisfied, and 0 otherwise. As shown in Figure 5.4, the optimal approach always achieves a satisfaction index of 1. Overall, the suboptimal approach has an improved satisfaction index over the benchmark. This is mainly due to the improved achieved throughput of the suboptimal approach as compared with the benchmark. While the suboptimal approach and benchmark satisfy the minimum required data rates of the MTs, the suboptimal approach achieves a much higher throughput than the benchmark due to the CSI-based bandwidth allocation, which leads to a higher satisfaction index.

5.3 Green Multi-homing Uplink Resource Allocation for Video Calls

Consider now an uplink multi-homing video transmission from an MT [168]. In the absence of an appropriate energy management strategy, the MT can use up all its available energy, and hence, might drain its battery before call completion. As a result, an energy management strategy is required to ensure a sustainable video transmission, over different radio interfaces, for the call duration. However, this problem has been overlooked, so far, in the literature. A simple energy management sub-system can equally distribute the MT available energy over different time slots of the video call duration. Given the time-varying bandwidth availability and channel conditions over different time slots, using this uniform energy distribution will lead to inconsistent temporal fluctuations in the video quality. An appropriate energy management sub-system should use the MT energy in a way such that it can support a consistent video quality in the call duration over time varying bandwidth and channel conditions.

In this section, an energy management sub-system is presented for MTs to support a sustainable multi-homing video transmission in a fading channel over a target call duration in a heterogeneous wireless access medium [161]. The energy management sub-system is based on a two-stage approach. In the first stage, through video quality statistical guarantee, the MT can determine a target video quality lower bound that can be supported for a target call duration with a small outage probability. In the second stage, the MT adapts its energy consumption during the call, following a three-step framework to achieve at least the target video quality lower bound.

The video sequence is encoded into a bit stream using a layered/scalable video encoder. The layered representation of the video sequence is composed of a base layer and several enhancement layers [169]. Each video layer is periodically encoded using a group-of-picture (GoP) structure. Time is partitioned into time slots, $\mathcal{T} = \{1, 2, \ldots, T\}$, of equal duration τ, where $T = \lceil T_c/\tau \rceil$ and T_c denotes the call duration, which is a random variable. At every time slot, the MT has a new GoP, from different layers, ready for transmission. The time slot duration is determined based on the source encoding rate in frames per second (fps). Each time slot has F frames from different layers, $\mathcal{F} = \{1, 2, \ldots, F\}$, and each frame can be of I, P or B type. I frames are compressed versions of raw frames independent of other frames. P frames only refer to preceding I/P frames, while B frames can refer to both preceding and succeeding frames. The data within one time slot are encoded inter-dependently through motion estimation, while data belonging to different time slots are encoded independently [170]. A video frame has the following characteristics [170]:

- Size—Each frame f is encoded into packets and each packet contains data relative to at most one frame [171]. Frame f is fragmented into C_f packets, $C_f \in [1, C_{f,\max}]$, where $C_{f,\max}$ denotes the maximum allowable size for frame f at each GoP. The frame size (in numbers of video packets, C_f) is represented by an independent identically distributed (i.i.d.) random variable that follows a probability mass function (PMF) $f_{C_f}(c_f)$ [170]. The frame size across different GoPs follows the same PMF given the frame type (I, P or B). The PMF, $f_{C_f}(c_f)$, can be calculated for different video contents and frame types as in [172]. The frame size, C_f, for frames of I, P or B type is constant within one time slot and varies from one time slot to another. The packet size (in bits) for frame f is denoted by ϖ_f.

- Distortion Impact—Each frame, f, has a distortion impact value per packet, v_f. It represents the amount by which video distortion is reduced if this packet is received on time at the decoder side. The packet distortion impact value, v_f, for different video contents and frame types can be calculated as in [173].
- Delay Deadline—It represents the time by which the frame should be decoded at the destination, which is also known as the decoding time stamp [168]. Packets that belong to the same frame have the same delay deadline, which is denoted by ϑ_f. Since videos are encoded using a fixed number of fps within the same layer, the difference in the delay deadline between any two consecutive frames within the layer is constant [168]. The delay difference is given by $|\vartheta_{f+1} - \vartheta_f| = \Theta_{f+1,f}$. The transmission deadlines of all packets within a given GoP expire within τ.
- Dependence—Within each time slot, since some frames are encoded based on the prediction of other frames, there are dependencies among these frames. Hence, packet decoding of one frame depends on the successful decoding of packets from other frames. These dependencies among packets of different frames, within one time slot, are expressed using a directed acyclic graph (DAG) [170], as shown in Figure 5.5. Hence, each video packet k_f is said to have ancestors $\mathcal{A}_k^f$. Packets which belong to $\mathcal{A}_k^f \, \forall f \in \mathcal{F}$ have higher distortion impact and smaller delay deadline than packet k_f.

Consider an uplink live video transmission from an MT [168]. The MT is equipped with multiple radio interfaces and presents multi-homing capabilities. As a result, the MT can establish communications with multiple wireless networks simultaneously and employ them for video packet transmission. Let $\mathcal{L} = \{1, 2, \dots, L\}$ denote the utilized radio interfaces and assume $L \geq 2$.

The uplink bandwidth allocated to the MT for radio interface l is denoted by b_l [4–6]. The offered bandwidth to the MT varies according to call arrivals and departures. Call arrivals follow a Poisson process, the channel holding time follows a general distribution and all calls are served without queueing. Hence, an $M/G/\infty$ model can be used to capture the statistics of the number of calls that are simultaneously in service [5, 6]. Hence, using the statistics of the

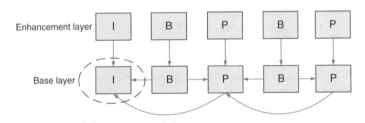

Figure 5.5 GoP structure with frame dependencies [161]. For instance, the circled I frame is an ancestor for the first B and P frames in the base layer and the I frame in the enhancement layer.

number of calls in service and the resource allocation mechanism, the probability that bandwidths $b_1, b_2, \ldots, b_L$ are offered to radio interfaces $1, 2, \ldots, L$, $f_{B_1, B_2, \ldots, B_L}(b_1, b_2, \ldots, b_L)$, can be derived based on a Poisson distribution [5, 6, 161].

The average transmission power allocated to radio interface l is denoted by $\bar{P}_l$. Let γ_l denote the received signal-to-noise ratio (SNR) at the BS/AP communicating with radio interface l. It is assumed that the channel conditions do not change much during one time slot. Hence, the received SNR value, $\gamma_l, \forall l \in \mathcal{L}$, is constant within one time slot and varies independently from one time slot to another [168, 170, 174].

Each radio interface, $l \in \mathcal{L}$, can support a discrete set of data rates r_{l,g_l}, with $g_l \in \mathcal{G} = \{1, 2, \ldots, G\}$. Radio interface $l \in \mathcal{L}$ can support data rate r_{l,g_l} if the received SNR value, γ_l, for this radio interface exceeds some threshold ζ_{l,g_l}. The set of thresholds $\zeta_{l,g_l}, \forall l \in \mathcal{L}, g_l \in \mathcal{G}$, can be calculated using Shannon formula as

$$\zeta_{l,g_l} = 2^{\frac{r_{l,g_l}}{b_l}} - 1, \quad l \in \mathcal{L}, g_l \in \mathcal{G}, \tag{5.16}$$

and $\zeta_{n,G+1}$ is assumed to be ∞.

For each time slot, let x_{kl}^f denote a video packet scheduling decision, where $x_{kl}^f = 1$ if packet k of frame f is assigned to radio interface l, otherwise $x_{kl}^f = 0$. Variable P_l denotes the instantaneous transmission power allocation to radio interface l. The circuit power required to keep radio interface l active is denoted by P_l^c. The MT available energy at the beginning of the call is denoted by E.

5.3.1   Energy Management Sub-system Design

The energy management sub-system consists of two stages. The first stage takes place during call set-up and aims to determine an optimal QoS lower bound that can be supported over the call duration, given the MT available energy, target call duration and video and radio interface characteristics. The second stage takes place during the call where the MT adapts its energy consumption to satisfy at least the target video quality lower bound calculated in the call set-up.

5.3.1.1   Statistical QoS Guarantee for Wireless Multi-homing Video Transmission

Let Q_t denote the video quality metric which is defined as the distortion impact ratio of the transmitted packets to the total available packets in time slot t. Due to channel fading and time-varying radio access bandwidth (and hence, time-varying data rates at different radio interfaces) and packet encoding statistics, the video quality metric Q_t is a discrete random variable. For a stationary and ergodic process underlying the system dynamics (in terms of channel fading, offered bandwidth and packet encoding), the time subscript t of Q_t can be omitted. Hence, Q is expressed as

$$Q = \frac{\sum_{k_f, f \in \mathcal{F}} \sum_{l \in \mathcal{L}} x_{kl}^f v_f}{\sum_{k_f, f \in \mathcal{F}} v_f}. \tag{5.17}$$

We aim to find the video quality cumulative distribution function (CDF), $F_Q(q)$, given the MT available energy, the time varying offered bandwidth and channel conditions at different radio interfaces, the target call duration and the video packet characteristics in terms of distortion impact, delay deadlines and packet encoding statistics. Using the video quality CDF, we can find the video quality lower bound, $\hat{q}$, that can be supported by the MT for the target call duration such that $\Pr(Q \leq \hat{q}) \leq \epsilon_q$, with $\epsilon_q \in [0, 1]$. This is achieved following a three-step framework:

1. *Calculating the probability of employing a given set of data rates at different radio interfaces:* In a Rayleigh fading channel, the probability that data rates $r_{1,g_1}, r_{2,g_2}, \ldots, r_{L,g_L}$ are used at radio interfaces $1, 2, \ldots, L$ can be calculated as [161]

$$
\begin{aligned}
& f_{R_{1,g_1}, \ldots, R_{L,g_L}}(r_{1,g_1}, \ldots, r_{L,g_L}) \\
& = \sum_{\mathcal{B}} \prod_{l=1}^{L} \left(\exp\left(-\frac{\zeta_{l,g_l}}{\bar{\gamma}_l} \right) - \exp\left(-\frac{\zeta_{l,g_l+1}}{\bar{\gamma}_l} \right) \right) \cdot f_{B_1, \ldots, B_L}(b_1, \ldots, b_L),
\end{aligned}
\tag{5.18}
$$

   where $\mathcal{B}$ denote the set of offered bandwidths to the MT, $\bar{\gamma}_l = (\bar{P}_l \Omega_l)/(b_l N_0)$ represents the average received SNR for radio interface l, Ω_l denotes the average channel power gain for radio interface l and N_0 denotes the one-sided noise power spectral density.
2. *Calculating the video quality CDF given the frame size and data rate statistics:* Next, we aim to find the video quality q that can be achieved given the MT data rates r_{l,g_l} at different radio interfaces and frame size c_f, where f can be of I, P or B type. Using the data rate and packet encoding statistics, we find the video quality CDF, $F_Q(q)$.

Since video packets that belong to the same frame have the same delay deadline of the frame, the required rate to transmit a packet $k_f, \forall f \in \mathcal{F}$, is given by $r(k_f) = \varpi_f / \Theta_{f+1,f}$ [168]. The scheduled packets to a given radio interface, l, should satisfy

$$
\sum_{k_f, f \in \mathcal{F}} x_{kl}^f r(k_f) \leq r_{l,g_l}, \quad \forall l \in \mathcal{L}, g_l \in \mathcal{G}.
\tag{5.19}
$$

Video packet scheduling should capture the dependence relationship among different video packets within the same time slot. The packets with unscheduled ancestors should not be transmitted since they will not be successfully decoded at the destination, and thus, they waste the MT and network resources. This requirement can be expressed by a precedence constraint

$$
x_{kl}^f \leq x_{k'l'}^{f'}, \quad \forall k_{f'}' \in A_k^f, l, l' \in \mathcal{L}.
\tag{5.20}
$$

Finally, a video packet should be assigned to one and only one radio interface

$$
\sum_{l=1}^{L} x_{kl}^f \leq 1, \quad \forall k_f, f \in \mathcal{F}.
\tag{5.21}
$$

Hence, multi-homing video packet scheduling, given the available data rates $r_{1,g_1}, r_{2,g_2}, \ldots, r_{L,g_L}$ at different radio interfaces and frame size c_f with f belonging to I, P or B type, should satisfy

$$
\begin{aligned}
& \max_{x_{kn}^f} \quad q \\
& \text{s.t.} \quad (5.19)\text{--}(5.21) \\
& \qquad x_{kn}^f \in \{0, 1\}.
\end{aligned}
\tag{5.22}
$$

The optimization problem (5.22) is a binary program. Problem (5.22) can be mapped to a new variant of the knapsack problem, referred to as precedence-constrained multiple knapsack problem (PC-MKP) [25]. The available items are the video packets, k_f, $f \in \mathcal{F}$, the item weights are the required data rates, $r(k_f)$ and the profit associated with each item is the packet distortion impact value, v_f. As we have multiple radio interfaces, the problem has multiple knapsacks each with capacity r_{l,g_l}. Due to the dependencies among different video packets within the time slot, MKP admits the precedence constraint (5.20). Since the knapsack problems are NP-hard [175], PC-MKP is also NP-hard. We present a greedy algorithm that can solve the PC-MKP of (5.22) in polynomial time based on [176]. Video packets are first classified into root and leaf items. In general, root items have higher precedence order than leaf items. For a video packet transmission, root items (packets of I and P frames) have higher distortion impact than leaf items (packets of B frames) [170]. Consider the following variables $\mathcal{U}$-the set of unassigned packets, u_l-the current used capacity at radio interface l (the remaining capacity is $o_l = r_{l,g_l} - u_l$), $\tilde{\mathcal{U}}_l$-the set of assigned packets to radio interface l ($\tilde{\mathcal{U}} = \overset{L}{\underset{l=1}{\cup}} \tilde{\mathcal{U}}_l$) and i_{k_f}-the index of the radio interface where packet k_f is currently assigned to. The multi-homing video packet scheduling algorithm is described in Algorithm 5.3.8.

Algorithm 5.3.8 consists of two parts. The first part (A1) aims to find a feasible solution for the problem by assigning items (video packets) with the highest profit (distortion impact) to different knapsacks (radio interfaces) while considering their precedence constraints. The second part (A2) aims to improve the feasible solution of (A1). This is achieved by considering all pairs of packed items (video packets) and, if possible, interchanges them whenever doing so allows the insertion of an additional item (video packet) from the remaining ones, if all its ancestors are packed, into one of the knapsacks (radio interfaces). In (A2) of Algorithm 5.3.8, $\tilde{\mathcal{U}},\mathcal{U}, o_l$ and i_{kf} are updated whenever some $\tilde{\mathcal{U}}_l$ is updated. If the total number of available video packets in a given time slot is $\sum_{f\in\mathcal{F}}c_f$, then the complexity of Algorithm 5.3.8 is $O(\sum_{f\in\mathcal{F}}c_f L) + O(\{\sum_{f\in\mathcal{F}}c_f\}^2)$, that is, has polynomial time complexity in terms of the number of radio interfaces and video packets.

Algorithm 5.3.8 Multi-Homing Video Packet Scheduling

A1: Finding a Feasible Solution

Input: r_{l,g_l} $\forall l \in \mathcal{L}, c_f \forall f \in \mathcal{F}$;

Initialization: $\mathcal{U} \longleftarrow \underset{f\in\mathcal{F}}{\cup} k_f, u_l \longleftarrow 0, \tilde{\mathcal{U}}_l = \{\}$ $\forall l \in \mathcal{L}$;

for $l \in \mathcal{L}$ **do**

    for $k_f \in \mathcal{U}$ **do**

        if $x^{f'}_{k'l'} = 1$ $\forall k'_{f'} \in \mathcal{A}^f_k, l' \in \mathcal{L}, r(k_f) + u_l \le r_{l,g_l}$ **then**

            $x^f_{kl} = 1, u_l = u_l + r(k_f)$;

        end if

        $\tilde{\mathcal{U}}_l = \tilde{\mathcal{U}}_l \cup \{k_f\}$;

    end for

    $\mathcal{U} = \mathcal{U} - \tilde{\mathcal{U}}_l$;

end for

for $l \in \mathcal{L}$ and $o_l > \min\{r(k_f)|k_f \in \mathcal{U}\}$ **do**

    for $k_f \in \mathcal{U}$ **do**

if $x_{k'_{f'}l'}^{f'} = 1 \, \forall k'_{f'} \in \mathcal{A}_k^f, l' \in \mathcal{A}, r(k_f) + u_l \leq r_{l,g_l}$ **then**
$\qquad x_{kl}^f = 1, u_l = u_l + r(k_f);$
end if
$\tilde{\mathcal{U}}_l = \tilde{\mathcal{U}}_l \cup \{k_f\};$
end for
$\mathcal{U} = \mathcal{U} - \tilde{\mathcal{U}}_l;$
end for
A2: Improving the Feasible Solution
for $k1 \in \{k_f | k_f \in \tilde{\mathcal{U}}, o_{i_{kf}} + \max\limits_{l \neq i_{kf}} o_l \geq \min\limits_{k'_{f'} \in \mathcal{U}} r(k'_{f'})\}$ **do**

$\qquad$ **for** $k2 \in \{k_f | k_f \in \tilde{\mathcal{U}}, k_f > k1, i_{kf} \neq i_{k1}, o_{i_{kf}} + o_{i_{k1}} \geq \min\limits_{k'_{f'} \in \mathcal{U}} r(k'_{f'})\}$ **do**

$\qquad\qquad W(a) = \max\{r(k1), r(k2)\}, W(b) = \min\{r(k1), r(k2)\};$
$\qquad\qquad j_a = i_a, j_b = i_b, \Delta = W(a) - W(b);$
$\qquad\qquad$ **if** $\Delta \leq o_{j_b}$ and $o_{j_a} + \Delta \geq \min\limits_{k'_{f'} \in \mathcal{U}} r(k'_{f'})$ **then**

$\qquad\qquad\qquad v_c = \max\{v_{k'_{f'}} | k'_{f'} \in \mathcal{U}, r(k'_{f'}) \leq o_{j_a} + \Delta, \mathcal{A}_{k'}^{f'} \subset \tilde{\mathcal{U}}\};$
$\qquad\qquad\qquad \tilde{\mathcal{U}}_{j_a} = (\tilde{\mathcal{U}}_{j_a} - a) \cup \{b, c\}, \tilde{\mathcal{U}}_{j_b} = (\tilde{\mathcal{U}}_{j_b} - b) \cup \{a\};$
$\qquad\qquad$ **end if**
$\qquad$ **end for**
end for
Output: $q = \dfrac{\sum_{k_f, f \in \mathcal{F}} \sum_{l \in \mathcal{L}} x_{kn}^f v_f}{\sum_{k_f, f \in \mathcal{F}} v_f}.$

Using Algorithm 5.3.8, the video quality q that can be achieved using data rates $r_{1,g_1}, r_{2,g_2}, \ldots, r_{L,g_L}$ at radio interfaces $1, 2, \ldots, L$ and frame size c_f with f belonging to I, B or P type can be calculated. The set of different data rates and packet encoding combinations that result in the same video quality q is denoted by $\mathcal{Q}$. We can map the data rate and frame size statistics into a video quality PMF given by

$$f_Q(q) = \sum_{\mathcal{Q}} \{f_{R_{1,g_1}, \ldots, R_{L,g_L}}(r_{1,g_1}, \ldots, r_{L,g_L}) \cdot f_{C_I, C_B, C_P}(c_I, c_B, c_P)\}, \qquad (5.23)$$

where $f_{C_I, C_B, C_P}(c_I, c_B, c_P)$ denotes the joint PMF of video packet encoding for I, B and P frames and it is obtained through the multiplication of the PMFs of I, B and P frames assuming an i.i.d. frame size statistics [170]. Consequently, the video quality CDF, $F_Q(q)$, can be calculated.

3) *Finding the maximum video quality lower bound $\hat{q}$ that can be supported for the target call duration:* From (5.18), the probability that data rates r_{l,g_l} are used at different radio interfaces depends on the average received SNR values $\bar{\gamma}_l, \forall l \in \mathcal{L}$. Therefore, the video quality CDF is a function of the average transmission power at different radio interfaces. Hence, the distribution of the average transmission power, E/T_c, among different radio interfaces, that is, $\bar{P}_l$, affects the resulting video quality CDF.

Since T_c is a random variable, we aim to guarantee that the MT available energy can support a target call duration, $\tilde{T}_c$. Hence, we first find $\tilde{T}_c$ that satisfies $\Pr(T_c \leq \tilde{T}_c) \geq 1 - \epsilon_c$, $\epsilon_c \in [0, 1]$. Assuming an ergodic process for system dynamics, in order to find the maximum video quality lower bound $\hat{q}$ that can be supported for the target call duration $\tilde{T}_c$ with some statistical guarantee ϵ_q, we need to solve

$$
\begin{aligned}
&\max_{\bar{P}_l \geq 0} \quad \hat{q} \\
&\text{s.t.} \quad F_Q(\hat{q}) \leq \epsilon_q \\
&\qquad \sum_{n=1}^{N}(\bar{P}_l + P_l^c) \leq \frac{E}{T_c}.
\end{aligned}
\tag{5.24}
$$

The first constraint in (5.24) represents an inequality (instead of an equality) condition since the supported data rates at different radio interfaces form a discrete set, and hence, the achieved video quality is also discrete. Consequently, an equality in the first constraint of (5.24) cannot always be satisfied, unlike the inequality. In (5.24), ϵ_q is a design parameter that can be chosen to strike a balance between the desired performance (in terms of the video quality and energy consumption) and success probability of the call delivery. This issue is further investigated in the performance evaluation subsection. The second constraint stands for the average power consumption of the MT, which is based on the total available energy and the target call duration. In the proposed energy management sub-system, the MT cannot assume an average energy consumption greater than that value.

Heuristic optimization techniques, for example, the genetic algorithm (GA) [177], can be used to solve the optimization problem (5.24). The GA can be easily implemented in smart phones as it consists of simple iterations. In addition, using the GA in solving (5.24) is fast due to the small number of variables (the number of radio interfaces ranges from 2 to 4).

Following (5.24), the MT can support a multi-homing video quality at least equal to $\hat{q}$ for the call duration T_c with an outage probability ϵ_s, given by

$$
\begin{aligned}
\epsilon_s &= 1 - \Pr(Q \geq \hat{q} | T_c \leq \tilde{T}_c) \cdot \Pr(T_c \leq \tilde{T}_c) \\
&= 1 - (1 - \epsilon_q) \cdot (1 - \epsilon_c).
\end{aligned}
\tag{5.25}
$$

5.3.1.2 Energy-Efficient QoS Provision for Wireless Multi-homing Video Transmission

During the call, the MT adapts its energy consumption to satisfy at least the maximum video quality lower bound $\hat{q}$ calculated in the call set-up. In good channel and/or network conditions, the MT achieves a video quality better than the lower bound. However, in bad conditions, the MT satisfies a quality not less than the lower bound. This strategy is performed in three steps:

Step 1: The MT determines the total required data rate, at the current time slot, to satisfy at least $\hat{q}$, given the current time slot video packet encoding. Let q_t denote the resulting video quality that can be achieved at time slot t by scheduling a set $\mathcal{U}$ of video packets for transmission. The total required data rate, r, can be calculated using Algorithm 5.3.9.

Algorithm 5.3.9 Calculation of Total Required Data Rate to Satisfy QoS Lower Bound

---
Input: $c_f \ \forall f \in \mathcal{F}$;

Initialization: $\mathcal{U} \longleftarrow \underset{f \in \mathcal{F}}{\cup} k_f, r \longleftarrow 0, \tilde{\mathcal{U}} = \{\}$;

while $q_t < \hat{q}$ **do**

    if $x^f_{k'_{f'} l'} = 1 \forall k'_{f'} \in \mathcal{A}^f_k, l' \in \mathcal{L}$ **then**

        $x^f_{kl} = 1, r = r + r(k_f)$;

    end if

    $\tilde{\mathcal{U}} = \tilde{\mathcal{U}} \cup \{k_f\}$;

end while

Output: r.

Step 2: The MT determines the minimum power required at each radio interface, and hence, the required data rate at each radio interface, to satisfy the total required data rate calculated in Step 1, given the current time slot channel fading and offered bandwidth. Let E_t denote the MT available energy at the beginning of time slot t. The transmission power allocation problem can be formulated as

$$\min_{P_l \geq 0} \quad \sum_{l=1}^{L} (P_l + P_l^c)\tau$$
$$\text{s.t.} \quad \sum_{l=1}^{L} r_{l,g_l} \geq r \tag{5.26}$$
$$\sum_{l=1}^{L} (P_l + P_l^c)\tau \leq E_t.$$

Similar to (5.24), (5.26) can be solved using the GA. Hence, during every time slot with duration τ, the MT updates its transmission power allocation P_l at each radio interface l to satisfy its target video quality.

Step 3: The MT performs video packet scheduling given the data rate at each radio interface, calculated in Step 2. Using the data rates r_{l,g_l} that can be supported through the transmission power allocation P_l calculated in (5.26), Algorithm 5.3.8 is used to schedule the current time slot available video packets for transmission. The resulting video quality satisfies the lower bound $\hat{q}$, calculated in (5.24), over the entire call duration with a success probability ϵ_s.

The energy management sub-system procedure for supporting a sustainable video transmission over the call duration with consistent video quality is summarized in Figure 5.6.

5.3.2 Performance Evaluation

The performance of the energy management sub-system, which is referred to as statistical guarantee framework (SGF), is compared with two benchmarks. The first benchmark, which

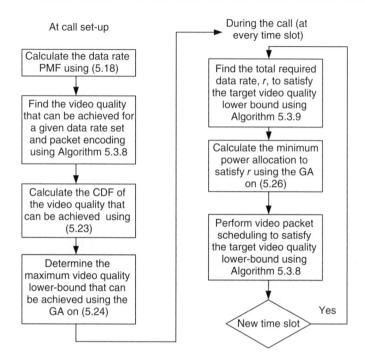

Figure 5.6   Flow chart of the proposed energy management sub-system procedure.

is referred to as total energy framework (TEF), aims to maximize the resulting video quality subject to the MT battery energy limitation [161]. The second benchmark, which is referred to as the equal-energy framework (EEF), satisfies an energy budget per time slot for energy management. A uniform energy budget per time slot is considered, where the MT available energy at time slot t is uniformly distributed over the remaining time slots [161].

Video sequences are compressed at an encoding rate of 30 fps [171]. The GoP structure consists of 13 frames with one layer (base layer) and one B frame between P frames. As a result, the time slot duration τ is 433 ms. The PMFs of the I, B and P frame sizes are given in [161]. The decoder time stamp difference between two successive frames, Θ, is 40 ms [168]. Each video packet requires a transmission data rate of 2 Kbps. The video packet distortion impact values are $v_f = 5$ for I frames, $v_f = 4$ for P frames and $v_f = 2$ for B frames [171]. Two radio interfaces are used for video transmission ($L = 2$). The circuit power for each radio interface is 10 mW. The offered bandwidth statistics on the two radio interfaces is given in [161]. The set of data rates that can be supported on each radio interface is $\mathcal{R} = \{0, 0.256, 0.512, 1, 1.5, 2, 2.5\}$ Mbps. Each radio interface is subject to a Rayleigh fading channel with average channel power gain $\Omega_1 = 0.5031$ and $\Omega_2 = 0.4852$. For comparison, a video call is established using the three set-ups SGF, TEF and EEF. The available energy at the beginning of the call for the three set-ups is 3 kJ. For the SGF, the video quality lower bound $\hat{q}$ is calculated in the call set-up, and it is equal to 89%, while $\epsilon_q = 0.1$ and $\epsilon_c = 0.3$.

Figure 5.7 plots the achieved video quality over the call duration using EEF, SGF and TEF. The TEF uses all the MT available energy, and hence it drains its battery before call completion.

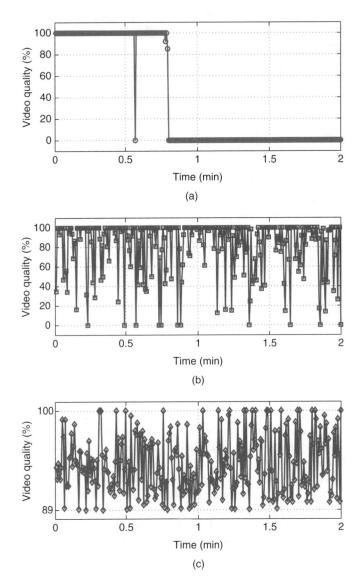

Figure 5.7 Performance comparison for the achieved video quality versus time using TEF, EEF and SGF [161]. $E = 3\,\text{kJ}$, $\epsilon_q = 0.1$ and $\epsilon_c = 0.3$.

This is because the main objective of TEF is to maximize the video quality in the current time slot, without considering the impact of the consumed energy on the video quality in the remaining time slots. The EEF takes into consideration the target call duration by equally distributing the MT available energy over the remaining time slots. However, due to the time-varying video packet encoding, offered bandwidths and channel conditions at the different radio interfaces, using the uniform energy budgets leads to inconsistent temporal fluctuations in the video quality. The resulting video quality for some time slots can be 0% as shown in Figure 5.7. On the

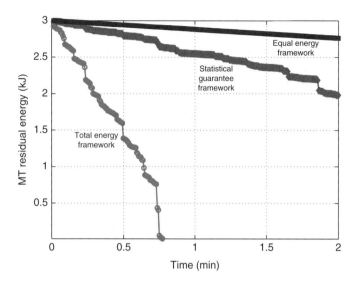

Figure 5.8   MT residual energy versus time. $E = 3$ kJ, $\epsilon_q = 0.1$ and $\epsilon_c = 0.3$.

contrary, the SGF can adapt the MT consumed energy at every time slot according to the packet encoding, offered bandwidth and channel conditions at the two radio interfaces. Consequently, the SGF can support a consistent video quality over different time slots, which is at least equal to the target lower bound (89%).

Figure 5.8 plots the MT residual energy over the call duration. The MT residual energy using the TEF near the middle of the call is insufficient to support a video transmission. Since the EEF uses a uniform energy budget for different time slots regardless of the channel fading, the slope of the consumed energy is almost constant over the first two-thirds of the call period. For the SGF, the MT consumed energy does not have an equal slope as the MT adapts its energy consumption based on the video packet encoding and channel conditions at the different radio interfaces over the time slot.

The advantages of SGF over the two benchmarks can be summarized as follows: (i) SGF guarantees a sustainable multi-homing video transmission over the target call duration, unlike TEF and (ii) SGF supports a consistent video quality over different time slots by adapting its energy consumption according to the video packet encoding and channel conditions at different radio interfaces, and consequently, SGF can control the QoS lower bound violation probability.

5.4   Summary

In this chapter, two radio resource mechanisms are presented to support green uplink multi-homing communications. The first mechanism is for data calls and is based on a joint bandwidth and power allocation framework that maximizes energy efficiency in a heterogeneous wireless medium. MTs are subject to minimum required data rates. The optimal framework jointly allocates bandwidth among MTs from different BSs/APs and the transmission power to the radio interfaces of each MT to maximize the minimum energy efficiency in the heterogeneous network. A desirable feature of the proposed framework is that it can be implemented in a decentralized manner among BSs/APs of different networks and MTs. A suboptimal framework is also presented to reduce the associated signalling overhead and computational complexity. The second mechanism supports a sustainable multi-homing video transmission over the call duration in a heterogeneous wireless access medium. The proposed framework aims to satisfy a target video quality lower bound that is calculated in the call set-up, given the MT available energy at the beginning of the call, the time-varying bandwidth availability and channel conditions at different radio interfaces, the target call duration and the video packet characteristics in terms of distortion impact, delay deadlines and video packet encoding statistics. The proposed framework enables the MT to support a consistent video quality over the call duration with a certain outage probability ϵ_s, by adapting its energy consumption according to the video packet encoding, offered bandwidth and channel conditions at different radio interfaces. Using ϵ_s as a design parameter, the MT can strike a balance between the desired performance (in terms of video quality and energy consumption) and the success probability of the call delivery.

6

Radio Frequency and Visible Light Communication Internetworking

Heterogeneous wireless networks mainly rely on RF network integration, such as cellular networks and WLANs or macro- and femto-cell cooperation as discussed in the previous chapters, to enhance the users perceived service quality and improve the achieved energy efficiency. However, such RF networks suffer from spectrum congestion. As a result, new network technologies that present larger spectrum availability should be introduced. In this context, VLC is considered to be a promising network technology that can offer high data rates with almost no transmission power consumption. Yet, VLC suffers from some technical limitations (such as absence of line of sight (LoS) and infeasibility of uplink transmission) that motivate its integration with RF network technologies to enhance the overall network performance. This chapter discusses RF and VLC integration in heterogeneous networks. Several integration objectives are presented such as load balancing, throughput maximization and uplink data transmission. Then, this chapter focuses on RF and VLC internetworking for green (energy-efficient) communications. A radio resource allocation mechanism that can improve downlink energy efficiency in an integrated RF–VLC network is presented and the challenging issues are discussed.

6.1 Introduction

Using visible light for communications dates back to the early ages, when humans relied on beacon fires and lighthouses to convey messages [178]. In the 19th century, Arthur Aldis invented the signal lamp, which uses Morse code for data transmission [179]. In 1880, Alexander Graham Bell invented the photophone, a device that modulates light beams with human voice [178]. However, several technical limitations (e.g. the impact of natural obstacles such as fog and rain on the perceived signal quality) prevented Bell from pursuing his research. Recently, the invention of light-emitting diode (LED) has renewed the research in VLC. In 2004, data transmission was possible in the Nakagawa laboratory using LEDs and in turn this has raised a significant interest from both academia and industry in VLC [180].

Green Heterogeneous Wireless Networks, First Edition. Muhammad Ismail, Muhammad Zeeshan Shakir, Khalid A. Qaraqe and Erchin Serpedin.
© 2016 John Wiley & Sons, Ltd. Published 2016 by John Wiley & Sons, Ltd.

In general, optical wireless communications refer to data transmission using infrared (IR), ultraviolet (UV) and visible light communications [181]. VLC is different from IR and UV in the sense that the same energy used for illumination is used for communications [181]. In particular, visible light is emitted from a LED light bulb, when excited by a direct current, as a stream of photons. The light intensity is proportional to the direct current value. By modulating the direct current with the data signal, the light intensity varies, which can be detected by a photo diode at the receiver. The varying light intensity, however, is undetectable to the human eyes. Hence, VLC is a data communication technology that uses the frequency spectrum in the range of 384–789 THz (i.e. corresponding to wavelengths 380–780 nm).

VLC offers many attractive features that motivates its wide adoption [178]. For instance, the VLC spectrum is 10,000 times more than the RF spectrum. In turn, this enables the support of data-hungry applications that require high throughput. Furthermore, there are no safety or health concerns for VLC. In addition, VLC offers a secure way of communication as the VLC signal is confined to the illumination area and cannot travel through walls. More importantly, VLC presents an energy-efficient means of communication, since the same energy used for illumination is used for communication, which implicitly means that zero transmission power is used in VLC. However, VLC technology is challenged by several technical limitations [178, 181]. For instance, in VLC, data communication deteriorates significantly in the absence of LoS signal, which is not the case for RF communications. Furthermore, interference from ambient (sun) light can significantly reduce the received SNR, and hence it degrades the communication quality. Moreover, realizing an efficient uplink communication can be problematic in VLC as MTs cannot support an optical uplink due to issues related to device orientation and energy constraints [182].

The aforementioned advantages and limitations suggest that VLC is better exploited if integrated with the existing RF communication technology rather than competing with it. In particular, there have been a few research efforts that aim to design a heterogeneous network that integrates RF and VLC networks in a unified framework to make use of the benefits of each technology and overcome the associated drawbacks. In this chapter, we discuss RF and VLC network integration for green (energy-efficient) communications. Towards this end, we first present some fundamentals related to VLC, followed by the integration scenarios of RF and VLC networks for load balancing and throughput maximization. Then, we present a radio resource allocation mechanism [183] that exploits the RF and VLC radio resources to support green communications. Finally, future research challenges in designing such a green VLC–RF network are discussed.

6.2 VLC Fundamentals

This section covers some background information about VLC networks. This includes the VLC transceiver, VLC channel, interference issues in VLC and VLC–RF internetworking.

6.2.1 VLC Transceivers

A VLC transceiver is shown in Figure 6.1. In the following, we will describe the constitutive elements of the VLC transmitter and receiver.

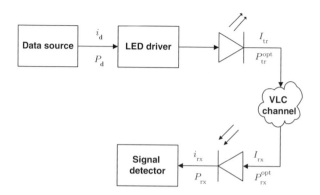

Figure 6.1 Illustration of VLC transceiver [183]

6.2.1.1 VLC Transmitter

From a lighting perspective, in practice, two types of LEDs can be used to generate the visible (white) light [181]. The first type is known as RGB LED, which mixes specific quantities of red, blue and green colours to generate the visible (white) light. However, this type of LED is challenged by inefficiency in green light generation (30% efficiency in generating the green light as compared with 80% and 60% in generating blue and red lights, respectively). Another more efficient type of LED is the phosphor-based white LED. This type of LED is commonly used in lighting. It generates blue light, converts part of the blue light and mixes the converted and non-converted parts of the blue light to generate the desired shade of white light.

From a communications perspective, denote i_{dr} as the driving current of the LED, expressed in Amperes. The driving current i_{dr} is proportional to the modulating data. Hence, the driving power P_{dr} is the average of the squared driving current, that is, $P_{dr} = i_{dr}^2$. In a multiuser system, MT m is allocated a proportion of P_{dr}, which is denoted by P_m. The allocated power P_m represents one system parameter to be controlled when optimizing the system performance. The LED output is an optical signal (visible light), whose intensity is proportional to the driving current and is expressed as $I_{tr} = \chi i_{dr}$, where χ is the proportionality factor of the electrical-to-optical conversion. The average optical transmitted power is given by $P_{tr}^{opt} = \bar{I}_{tr}$. The optical transmitted power specifies the LED illumination level.

The IEEE 802.15.7 standard, launched in September 2011, defines seven colour channels for the physical layer (PHY) in VLC [178]. Three PHY modes are introduced in IEEE 802.15.7, namely PHY I, PHY II and PHY III, which offer 11.67–266.6 kbps, 1.25–96 Mbps and 12–96 Mbps, respectively [184]. Both PHY I and PHY II are defined for a single light source and can support on–off keying (OOK) and variable pulse position modulation (VPPM). Only PHY III uses multiple optical sources with different frequencies (colours) and supports colour shift keying (CSK). Each PHY mode identifies mechanisms for light source modulation, run length-limited line (RLL) coding and channel coding. In addition, the achievable data rates can be significantly improved if more advanced modulation schemes are used such as optical orthogonal frequency division modulation (OOFDM) [180]. High data rates in the order of Gbps have also been achieved using multiple-input multiple-output (MIMO) techniques [180].

6.2.1.2 VLC Receiver

Illuminance is the most significant parameter that characterizes the white LEDs for lighting purposes [185]. Illuminance expresses the brightness of an illuminated surface [186]. At a specific point in the receiver plane, and assuming a Lambertian radiation pattern, the horizontal illuminance, as sensed by a photo-detector, is expressed as [185, 186]

$$I_{\text{rx}} = I_0 \frac{\cos^\omega(\phi) \cos(\theta)}{d^2}, \tag{6.1}$$

where I_0 denotes the maximal (center) luminous intensity, ϕ denotes the irradiance angle, θ is the incidence angle and d is the distance to the illuminated surface (i.e. the distance from LED to photo-detector). The Lambertian emission order ω is expressed as $-1/\log_2(\cos(\phi_{1/2}))$, where $\phi_{1/2}$ denotes the semi-angle at half-power (the viewing angle).

From a communication perspective, the received optical signal is first gathered by a concentrator. Different concentrators are used to gather LoS and non-LoS (NLoS) signals [181]. Optical filtering is then applied to alleviate the effect of the ambient light interference. The filtered signal is then fed to a photo-detector that converts the optical signal into an electrical signal (photo-current). Two types of photo-detectors can be used, namely photo-diode and image sensor [181]. The photo-current is then amplified, equalized and high pass filtered to remove the DC signal.

Denote the VLC channel power gain for MT m by h_m. The received optical signal by MT m is given by $I_{\text{rx},m} = h_m I_{\text{tr},m}$. Let ς represent the photo-detector responsivity (in amp/W), which denotes the efficiency of the photo-detector in converting the received optical intensity into an electrical current. Hence, the received electrical signal, $i_{\text{rx},m}$, for MT m is given by $i_{\text{rx},m} = \varsigma I_{\text{rx},m}$. The average electrical power of the received signal is given by $P_{\text{rx},m} = i_{\text{rx},m}^2$, and hence, it turns out that [183]

$$P_{\text{rx},m} = (\chi \varsigma h_m)^2 P_m. \tag{6.2}$$

6.2.2 VLC Channel

In the literature, two VLC channel power gain models are adopted. The first model captures only the LoS component, while the second model captures both the LoS component and NLoS first reflection. The two models are discussed below.

6.2.2.1 LoS Model

In [180], it has been shown that the average optical power received from the first reflection accounts only for 3.5% of the optical power received from the LoS path. Hence, several works in the literature such as [185–189] and [190] represent the VLC channel via the LoS path and discard the reflection contributions. The optical channel gain for the LoS signal (DC gain) as received by MT m is given by

$$H_m^{\text{LoS}} = \begin{cases} \dfrac{(\omega_m + 1)A_m}{2\pi d_m^2} F(\theta_m) G(\theta_m) \cos^{\omega_m}(\phi_m) \cos(\theta_m) & \theta_m \leq \Theta_m \\ 0 & \theta_m > \Theta_m, \end{cases} \tag{6.3}$$

where A_m is the physical area of the photo-detector of MT m, Θ_m is the half-angle of MT m field of view (FoV), $F(\theta_m)$ is the optical filter gain and $G(\theta_m)$ is the concentrator gain, which can be determined as

$$G(\theta_m) = \begin{cases} \dfrac{\epsilon^2}{\sin^2 \Theta_m} & 0 \le \theta_m \le \Theta_m \\ 0 & \theta_m > \Theta_m, \end{cases} \tag{6.4}$$

where ϵ denotes the refractive index.

However, most of the existing research do not account for the fact that the LoS component can be blocked due to obstacles. In [191], a random variable κ is introduced in (6.4) to denote the VLC LoS blocking event. The random variable κ is assumed to follow a Bernoulli distribution with blocking probability p.

6.2.2.2 NLoS Model

Although the NLoS component accounts for a small percentage of the optical power as compared with the LoS component, it is important in some scenarios such as when the LoS component is blocked and only the NLoS component is present or to account for the interference power from another AP. Therefore, it is important to model the NLoS channel power gain. In [191] and [192], the NLoS DC gain after the first diffuse reflection from a wall is expressed as

$$\mathrm{dh}_m = \begin{cases} \epsilon \dfrac{(\omega_m + 1) A_m \mathrm{dA}_{\text{wall}}}{4\pi^2 d_{1,m}^2 d_{2,m}^2} \cos(\psi_{1,m}) \cos(\psi_{2,m}) Z(\theta_m, \phi_m) & 0 \le \theta_m \le \Theta_m \\ 0 & \theta_m > \Theta_m, \end{cases} \tag{6.5}$$

where $d_{1,m}$ and $d_{2,m}$ denote the distance between the AP and a reflecting surface and the distance between the reflecting surface and MT, respectively, ϵ denotes the wall reflection index, $\mathrm{dA}_{\text{wall}}$ is a small reflective area, $\psi_{1,m}$ and $\psi_{2,m}$ denote the angle of irradiance to a reflective point and the angle of irradiance to the MT, respectively and $Z(\theta_m, \phi_m) = F(\theta_m) G(\theta_m) \cos^{\omega_m}(\phi_m) \cos(\theta_m)$. The NLoS channel power gain is then evaluated as follows

$$h_m^{\text{NLoS}} = \int_{\text{wall}} \mathrm{dh}_m, \tag{6.6}$$

where the integration is over the four wall areas.

The channel power gain h_m includes both an LoS component, h_m^{LoS}, and NLoS component, h_m^{NLoS}, when the AP is in the FoV of the MT; otherwise, only the NLoS component is considered [192] (which can be the case for an interferer).

6.2.2.3 Noise Contribution

Two noise components should be accounted for, namely the shot noise due to the ambient light and thermal noise [191]. Hence, the VLC Gaussian noise has a total variance of

$$\sigma_m^2 = \sigma_{\text{shot},m}^2 + \sigma_{\text{thermal},m}^2. \tag{6.7}$$

The shot noise variance is given by Jin et al. [191]

$$\sigma_{\text{shot},m}^2 = 2q\beta P_{\text{rx},m}^{\text{opt}} B_{\text{eq},m} + 2q I_{\text{bg}} I_1 B_{\text{eq},m}, \tag{6.8}$$

where q is the electronic charge, $P^{\text{opt}}_{\text{rx},m}$ is the received optical power by MT m, $B_{\text{eq},m}$ denotes the equivalent noise bandwidth, I_{bg} denotes the background current due to the background (ambient) light and I_1 is a constant (in [191] $I_1 = 0.562$). The thermal noise variance is given by Jin et al. [191]

$$\sigma^2_{\text{thermal},m} = \frac{8\pi\varrho T}{V}\Lambda_m A_m I_1 B^2_{\text{eq},m} + \frac{16\pi^2\varrho T\Gamma}{\varsigma_m}\Lambda^2_m A^2_m I_2 B^3_{\text{eq},m}, \tag{6.9}$$

where ϱ denotes the Boltzmann's constant, T is the absolute temperature, V denotes the open-loop voltage gain, Λ_m represents the fixed capacitance of the photo-detector per unit area, Γ is the FET channel noise factor, ς_m is the FET trans-conductance and I_2 is a constant (in [191] $I_2 = 0.0868$).

6.2.3 Interference Issues in VLC

In VLC, a single AP is a collection of LEDs installed in the ceiling of a room. The AP purpose is twofold: illumination and communication. The AP coverage area is defined as an area free of illumination and communication dead zones. The illumination dead zone presents an illuminance level below a target value (e.g. for an indoor environment, the desired illuminance level is 100–500 lumens as standardized by GB 50034-2004 in China [186]). Similarly, the communication dead zone presents an SINR value below a target threshold. The AP illumination and communication coverage area depends on several parameters such as the LEDs semi-angle at half-power, height of the LEDs and FoV of the receiver (MT) [186].

A VLC cell can be defined within the VLC AP coverage area. Unity frequency reuse is adopted when the same frequency f is reused across all VLC cells in a system with multiple APs. The drawback of such an approach is the resulting inter-cell interference (ICI) imposed by the neighbouring cells at the cell edge. For instance, the SINR for the cell-edge MT m given the cell formation shown in the top left area of Figure 6.2 [188] is given by

$$\gamma_m = \frac{P^A_{\text{rx},m}}{\sigma^2_m + P^B_{\text{rx},m} + P^C_{\text{rx},m} + P^D_{\text{rx},m}}, \tag{6.10}$$

where $P^A_{\text{rx},m}$ denotes the LoS component received by MT m from AP A; $P^B_{\text{rx},m}$, $P^C_{\text{rx},m}$ and $P^D_{\text{rx},m}$ denote the NLoS component (interference) received by MT m from APs B, C and D, respectively (assuming that only AP A is within the field of view of MT m) and σ^2 is given by (6.7). Hence, the MT receives interference from the NLoS component of all neighbouring cells. Different techniques are proposed in the literature to mitigate such ICI, as discussed below.

6.2.3.1 Non-Unity Frequency Reuse

In this technique, adjacent cells are allocated different frequency bands for communications. The top right area of Figure 6.2 represents cell formation with a frequency reuse factor of 2 [180, 188]. Hence, the SINR for cell-edge MT m is given by

$$\gamma_m = \frac{P^A_{\text{rx},m}}{\sigma^2_m + P^D_{\text{rx},m}}. \tag{6.11}$$

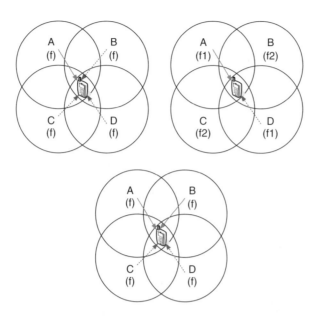

Figure 6.2   VLC interference in different cell formations [180, 188]

In (6.11), the ICI from cells B and C is eliminated as compared with (6.10). However, using a frequency reuse factor larger than 1, a trade-off exists between the bandwidth efficiency (which will be degraded) and the SINR for cell-edge users (which will be improved). Rather than allocating separate frequency bands for the entire cell, and hence reducing the overall bandwidth efficiency, power control and subcarrier allocation can be exploited to mitigate interference for cell-edge users. Such an approach is similar in concept to the dynamic fractional frequency reuse that is deployed in RF networks [192].

6.2.3.2 Combined Transmission

In a combined transmission technique, a group of neighbouring APs forms a cluster and transmits the same information to MT m [180, 188]. Hence, the received signals at the MT will be combined rather than being treated as interference. For instance, the bottom area of Figure 6.2 represents the combined transmission from two cells A and B to a cell-edge MT m. In this case, the SINR is given by

$$\gamma_m = \frac{P^A_{\text{rx},m} + P^B_{\text{rx},m}}{\sigma^2_m + P^C_{\text{rx},m} + P^D_{\text{rx},m}}. \tag{6.12}$$

In (6.12), the signal received from cell B is turned into a useful signal rather than being treated as interference as in (6.10), leading to an improved SINR. It should be noted that in (6.10) only the NLoS component of cell B contributes as an interference signal, while in (6.12) the LoS component is treated as a useful signal. One drawback with the combined transmission technique is that it also reduces the bandwidth efficiency, since only one user is served by several APs at a time.

Transmit pre-coding techniques (vector transmission) can be employed to enable simultaneous serving of multiple users [180, 188]. One example of such vector transmission is the zero-forcing approach, which is commonly used in multiuser MIMO RF communications.

6.2.4   VLC–RF Internetworking

As introduced earlier, VLC offers several advantages such as large available spectrum and energy-efficient and secure communications, while also it is challenged by several limitations such as signal deterioration in the absence of LoS and the infeasible uplink communications. Consequently, several works in the literature propose to integrate VLC and RF networks for improved system performance. In this context, VLC–RF internetworking has been employed for different objectives such as load balancing, throughput maximization, uplink support and energy efficiency, as discussed below.

6.2.4.1   Load Balancing

The load balancing problem is basically a joint user association and resource allocation problem. In the literature, different integration scenarios are considered with load balancing objectives such as the integration between VLC and RF femto cells as in [185] and VLC and WLAN in [188]. One goal of load balancing is to ensure that no network is congested with users in a way that deteriorates the users' perceived QoS. A logarithmic utility function is commonly used as it can achieve load balancing and fairness among mobile users [185, 188]. The user association is a binary decision variable x_{nm} that is set to 1 if MT m is assigned to network n (VLC or RF). A constraint over x_{nm} ensures that MT m is assigned to only one network (i.e. single-network assignment) [185, 188]. The radio resource allocation is a real variable that specifies the amount of allocated radio resources (e.g. power and/or bandwidth) to each MT with appropriate resource availability constraints. In most cases, the load balancing problem is an MINLP [188], which is an NP problem. In order to simplify the problem formulation, in [185], the authors consider that the effective load of an AP is represented by the number of MTs associated with that AP. In this case, the optimal resource allocation policy is to equally distribute the resources among the MTs associated with the AP. Hence, the load balancing problem is reduced to a pure association problem, that is, the problem is reduced from MINLP to binary programming (BP) and the problem is solved through variable relaxation. Furthermore, other heuristic association strategies can be employed such as the minimum distance user association and the maximum SINR user association. When both user association and resource allocation are considered, linear approximation can be used to reduce the problem from MINLP to linear programming (LP) as in [188].

6.2.4.2   Throughput Maximization

In VLC, the achievable throughput spatially fluctuates within the indoor environment due to the absence of LoS caused by obstacles. Hence, the integration of VLC and RF networks is expected to improve the overall throughput, since VLC can support high data rates in certain areas while RF networks can support moderate data rates in larger areas. Such an integration with the objective of throughput maximization is studied in [190] between 5G RF AP (based on mmwave communications) and VLC, and in [191] between 4G RF femto cell and VLC. In [190], a central entity is assumed to be in place to monitor the network. By default, MTs connect to the VLC AP, and when the achievable throughput for a given MT is below a target threshold, the central entity connects this MT to the RF AP. In addition to investigating the achievable throughput in such an integrated system, the study in [190] investigates the throughput outage. It has been shown that the integrated system results in an improved achieved throughput and outage performance. Unlike [190], the work in [191] enables MTs to connect to both VLC and RF APs in a multi-homing approach. Furthermore, the work in [191] applies the effective capacity concept to convert the statistical delay constraints into equivalent average rate constraints. The objective of [191] is to design a radio resource allocation algorithm that maximizes the effective capacity of the integrated system and satisfies the users' required QoS. The problem is shown to be a convex optimization one that can be solved in a decentralized manner following the Lagrangian decomposition approach. Simulation results have demonstrated that in the presence of a VLC LoS component, the VLC AP is more reliable in satisfying the statistical delay guarantee than the RF AP, while in the absence of VLC LoS component, the RF AP becomes imperative.

6.2.4.3 Uplink Support

VLC is ideal to support downlink communications in the coverage area of its AP. However, supporting uplink communications in VLC is challenging due to MT energy constraints, device orientation and interference. A heterogeneous RF–VLC network can resolve such a challenge where downlink communication is supported by VLC AP and uplink communication is supported by RF AP. Several implementation issues that should be addressed to achieve this vision are investigated in [182] and a test bed is developed as well.

6.2.4.4 Energy-Efficient Communications

LED is classified as a green lighting technology. Reports have indicated that the total lighting global power consumption will be reduced by 50% if all light sources were replaced by LEDs [181]. In addition, VLC based on LED technology offers a green communication approach. This is mainly due to the fact that VLC exploits the illumination energy, which is already consumed for lighting, in high data rate transmission. However, due to reliability issues caused by the absence of LoS signal, VLC networks must be complemented by RF networks for energy-efficient and reliable network operation. In this context, the VLC AP employs its illumination power for data transmission while consuming additional power for data processing, while the RF AP consumes both data processing and transmission powers. MTs with multi-homing capability can receive data from both VLC and RF APs leading to an improved energy efficiency and reliability in the network performance. However, several challenging issues should be addressed first to fully exploit the energy efficiency and reliability benefits of such an integrated network. These challenging issues together with a radio resource allocation mechanism are discussed in the next section.

6.3   Green RF–VLC Internetworking

In this section, we first investigate the energy efficiency performance of an integrated VLC–RF network, then we discuss the challenging issues that need to be addressed in future research to fully exploit the energy saving and reliability benefits of this integrated network.

Consider an indoor downlink scenario with a single RF AP and VLC AP and a set of MTs $\mathcal{M}$ in their coverage areas, as shown in Figure 6.3. The RF AP can be a femto-cell AP or WLAN AP. MTs with multi-homing capability can receive data from both VLC and RF APs to satisfy its required QoS. The total available bandwidths at the VLC and RF APs are denoted by B_{VLC} and B_{RF}, respectively. The received bandwidths by MT m from the VLC and RF APs are denoted by $B_{\mathrm{VLC},m}$ and $B_{\mathrm{RF},m}$, respectively. The allocated transmission powers to MT m by the VLC and RF APs are given by $P_{\mathrm{VLC},m}$ and $P_{\mathrm{RF},m}$, respectively. Each AP has a maximum allowed total transmission power P_{VLC} and P_{RF}. In addition to the transmission power, each AP has fixed power consumption components $P_{\mathrm{VLC}}^{\mathrm{c}}$ and $P_{\mathrm{RF}}^{\mathrm{c}}$ for circuit operation and data processing before transmission. Furthermore, the VLC fixed power consumption accounts also for any required additional power to compensate for the losses in the LED efficiency due to data transmission. The achieved data rates by MT m from each AP are denoted by $R_{\mathrm{VLC},m}$ and $R_{\mathrm{RF},m}$, respectively. The total data rate achieved by MT m should satisfy a minimum required data rate of R_m. The channel power gain between MT m and the VLC AP, $h_{\mathrm{VLC},m}$, is dominated by the LoS signal path loss and is given by (6.3). The noise power spectral density affecting the VLC receivers is denoted by $N_{0,\mathrm{VLC}}$, and is given by $\sigma_m^2/B_{\mathrm{VLC},m}$, where σ_m^2 is given by (6.7). The channel power gain between MT m and the RF AP is dominated by the signal path loss, and it is given by

$$\mathrm{PL[dB]} = C_1 \log_{10}(d_{\mathrm{RF},m}) + C_2 + C_3 \log_{10}\left(\frac{f_c}{5}\right) + C_5, \qquad (6.13)$$

where f_c denotes the carrier frequency in GHz; C_1, C_2 and C_3 are constants depending on the propagation model and X is an environment-specific term. For the LoS scenario, $C_1 = 18.7$, $C_2 = 46.8$ and $C_3 = 20$. For the NLoS scenario, $C_1 = 36.8$, $C_2 = 43.8$, $C_3 = 20$ and

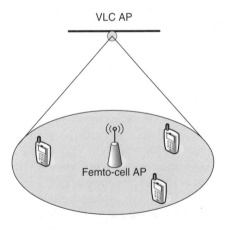

Figure 6.3   Illustration of VLC and RF APs coverage [183, 188]

$C_5 = 5(w - 1)$ in case of light walls or $C_5 = 12(w - 1)$ in case of heavy walls, where w denotes the number of walls between the AP and MT. Hence, the RF channel power gain is given by

$$h_{RF,m} = 10^{-PL[dB]/10}. \qquad (6.14)$$

Denote the RF channel power gain in the presence of LoS by $h_{RF,m}^{LoS}$ and for the NLoS scenario by $h_{RF,m}^{NLoS}$. The LoS availability probabilities for RF and VLC systems denote the probability that there are no obstacles in the communication link between the MT and the corresponding AP, and are denoted by κ_{RF} and κ_{VLC}, respectively. In the case of RF transmissions, the channel path-loss exponent increases with the LoS absence, as discussed earlier. For the case of VLC, the signal is degraded significantly in the absence of LoS and may result in unsuccessful data transmissions. It is assumed that the NLoS VLC transmissions are unsuccessful. Therefore, we focus only on the system performance with LoS VLC transmissions. This is mainly due to the fact that MTs operate in a multi-homing manner and the RF received signal will be significantly higher than the NLoS VLC signal.

6.3.1 Energy Efficiency Maximization

The average received electrical SNR for MT m from each AP is given by

$$\gamma_{RF,m} = \frac{P_{RF,m} h_{RF,m}}{B_{RF,m} N_{0,RF}}, \qquad (6.15)$$

$$\gamma_{VLC,m} = \frac{P_{VLC,m} (\chi\varsigma h_{VLC,m})^2}{B_{VLC,m} N_{0,VLC}}, \qquad (6.16)$$

where the value of $h_{RF,m}$ is given by $h_{RF,m}^{LoS}$ and $h_{RF,m}^{NLoS}$ to determine the corresponding $\gamma_{RF,m}^{LoS}$ and $\gamma_{RF,m}^{NLoS}$ for the LoS and NLoS RF channels, respectively. The value of $\gamma_{VLC,m}$ is calculated by dividing the received electrical power in (6.2) by the VLC noise power.

The achievable data rates from each AP are averaged over the probability mass function of LoS availability, and are given by

$$R_{RF,m} = B_{RF,m}(\kappa_{RF}\log_2(1 + \gamma_{RF,m}^{LoS}) + (1 - \kappa_{RF})\log_2(1 + \gamma_{RF,m}^{NLoS})), \qquad (6.17)$$

$$R_{VLC,m} = B_{VLC,m}\kappa_{VLC}\log_2(1 + \gamma_{VLC,m}). \qquad (6.18)$$

The total achieved data rate in the heterogeneous RF–VLC network is denoted by R_T, and assumes the expression

$$R_T = \sum_{m\in\mathcal{M}} R_{VLC,m} + \sum_{m\in\mathcal{M}} R_{RF,m}. \qquad (6.19)$$

The total communication power consumption is denoted by P_T and its value is calculated as follows

$$P_T = P_{VLC}^c + P_{RF}^c + \sum_{m\in\mathcal{M}} P_{RF,m}, \qquad (6.20)$$

where the first term in (6.20) represents the consumed power for the VLC AP and is calculated using the fact that the transmission power is the optical power used for illumination by design, and therefore only the fixed power consumption P_{VLC}^c is accounted for as a communication power cost. The second and third terms represent the RF power consumption, which accounts for both the processing and transmission powers.

The objective of the radio resource allocation mechanism is to allocate transmission powers and bandwidths for the MTs from the available APs such that the total energy efficiency of the heterogeneous network is maximized and the MTs target QoS is supported while the bandwidth and power limitations of the APs are satisfied. This optimization problem is formulated as follows

$$\max_{P_{\mathrm{VLC},m},P_{\mathrm{RF},m},B_{\mathrm{VLC},m},B_{\mathrm{RF},m}} \eta$$

$$\text{s.t.} \quad R_{\mathrm{VLC},m} + R_{\mathrm{RF},m} \geq R_m, \quad \forall m \in \mathcal{M},$$

$$\sum_{m \in \mathcal{M}} P_{\mathrm{VLC},m} \leq P_{\mathrm{VLC}},$$

$$\sum_{m \in \mathcal{M}} P_{\mathrm{RF},m} \leq P_{\mathrm{RF}},$$

$$\sum_{m \in \mathcal{M}} B_{\mathrm{VLC},m} \leq B_{\mathrm{VLC}},$$

$$\sum_{m \in \mathcal{M}} B_{\mathrm{RF},m} \leq B_{\mathrm{RF}},$$

$$P_{\mathrm{VLC},m}, P_{\mathrm{RF},m}, B_{\mathrm{VLC},m}, B_{\mathrm{RF},m} \geq 0, \quad \forall m \in \mathcal{M}, \qquad (6.21)$$

where $\eta = R_{\mathrm{T}}/P_{\mathrm{T}}$ denotes the heterogeneous network total energy efficiency. The radio resource allocation problem (6.21) is a concave–convex fractional program [164] since the numerator of the objective function is concave with respect to the decision variables and the denominator is affine.

As in Chapter 5, the fractional program in (6.21) can be converted into a convex optimization problem in terms of parameter λ. Considering the same constraints as in (6.21), define

$$F(\lambda) = \max_{P_{\mathrm{VLC},m},P_{\mathrm{RF},m},B_{\mathrm{VLC},m},B_{\mathrm{RF},m}} R_{\mathrm{T}} - \lambda P_{\mathrm{T}}. \qquad (6.22)$$

Using the Dinkelbach-type procedure in Algorithm 6.3.10, λ is updated and we can find the roots of $F(\lambda) = 0$, and hence the optimal solution for the optimization problem (6.22).

Algorithm 6.3.10 Dinkelbach-Type Procedure

Initialization: $P_{\mathrm{VLC},m}(0), P_{\mathrm{RF},m}(0), B_{\mathrm{VLC},m}(0), B_{\mathrm{RF},m}(0) > 0 \; \forall m, \lambda(1) = \eta(0), i = 1;$

while $F(\lambda(i)) \neq 0$ **do**

    Solve (6.22) for optimal$\{P_{\mathrm{VLC},m}(i), P_{\mathrm{RF},m}(i), B_{\mathrm{VLC},m}(i), B_{\mathrm{RF},m}(i)\} \forall m \in \mathcal{M};$

    $\lambda(i+1) = \eta(i);$

    $i \leftarrow i+1$

end while

Output: $P_{\mathrm{VLC},m}, P_{\mathrm{RF},m}, B_{\mathrm{VLC},m}, B_{\mathrm{RF},m} \forall m \in \mathcal{M}$

Problem (6.22) is a convex optimization problem. Applying the KKT conditions on the Lagrangian function of (6.22) as we did in Chapters 4 and 5, we can find (i) the optimal power allocations at the MT as a function of the Lagrangian multipliers μ_m, ν_{VLC} and ν_{RF} and (ii) the optimal bandwidth allocation at each AP as a function of the Lagrangian multipliers μ_m, β_{VLC} and β_{RF} [183]. The optimal values of the Lagrangian multipliers can be obtained by solving the dual problem using a gradient descent method [183].

Algorithm 6.3.11 finds the optimal VLC power allocation for a given allocated VLC bandwidth and λ.

Algorithm 6.3.11 Power Allocation for VLC Network

Input: $B_{\text{VLC},m}$, μ_m, and λ;

Initialization: $\nu_{\text{VLC}}(1) \geq 0$;

for $i = 1 : I$ **do**

    for $m \in \mathcal{M}$ **do**

$$P_{\text{VLC},m}(i) = B_{\text{VLC},m}\left[\kappa_{\text{VLC}}\frac{1+\mu_m}{(\nu_{\text{VLC}}(i)+\lambda)\ln(2)} - \frac{N_{0,\text{VLC}}}{(\chi_5 h_{\text{VLC},m})^2}\right]^+;$$

    end for

    $\nu_{\text{VLC}}(i+1) = \left[\nu_{\text{VLC}}(i) - \delta_1\left(P_{\text{VLC}} - \sum_{m\in\mathcal{M}} P_{\text{VLC},m}\right)\right]^+;$

end for

Output: $P_{\text{VLC},m} \forall m \in \mathcal{M}$

Similarly, Algorithm 6.3.12 finds the optimal RF power allocation for a given allocated RF bandwidth and λ.

Algorithm 6.3.12 Power Allocation for RF Network

Input: $B_{\text{RF},m}$, μ_m, and λ;

Initialization: $\beta_{\text{RF}}(1) \geq 0$;

for $i = 1 : I$ **do**

    for $m \in \mathcal{M}$ **do**

    Find $P_{\text{RF},m}(i)$ as the positivereal root of $\dfrac{B_{\text{RF},m}}{\ln 2}\left(\kappa_{\text{RF}}\dfrac{h_{\text{RF},m}^{\text{LOS}}}{P_{\text{RF},m}h_{\text{RF},m}^{\text{LOS}}+B_{\text{RF},m}N_{0,\text{RF}}} + \right.$

    $\left.(1-\kappa_{\text{RF}})\dfrac{h_{\text{RF},m}^{\text{NLOS}}}{P_{\text{RF},m}h_{\text{RF},m}^{\text{NLOS}}+B_{\text{RF},m}N_{0,\text{RF}}}\right) = \dfrac{\lambda+\nu_{\text{RF}}}{1+\mu_m};$

    end for

    $\nu_{\text{RF}}(i+1) = \left[\nu_{\text{RF}}(i) - \delta_2\left(P_{\text{RF}} - \sum_{m\in\mathcal{M}} P_{\text{RF},m}\right)\right]^+;$

end for

Output: $P_{\text{RF},m} \forall m \in \mathcal{M}$

Given the allocated VLC power from Algorithm 6.3.11, the optimal VLC bandwidth can be allocated using Algorithm 6.3.13.

Algorithm 6.3.13 Bandwidth Allocation for VLC Network

Input: $P_{\text{VLC},m}$, μ_m, and λ;

Initialization: $\beta_{\text{VLC}}(1) \geq 0$;

for $i = 1 : I$ **do**

for $m \in \mathcal{M}$ **do**

Find $B_{\text{VLC},m}(i)$ as the positive real root of $\kappa_{\text{VLC}} \left(\log_2 \left(1 + \frac{(\chi \varsigma h_{\text{VLC},m})^2 P_{\text{VLC},m}}{B_{\text{VLC},m} N_{0,\text{VLC}}} \right) \right.$

$$\left. - \frac{P_{\text{VLC},m} (\chi \varsigma h_{\text{VLC},m})^2}{\ln(2) \left(B_{\text{VLC},m} N_{0,\text{VLC}} + P_{\text{VLC},m} (\chi \varsigma h_{\text{VLC},m})^2 \right)} \right) = \frac{\beta_{\text{VLC}}}{1 + \mu_m};$$

end for

$$\beta_{\text{VLC}}(i+1) = \left[\beta_{\text{VLC}}(i) - \delta_3 (B_{\text{VLC}} - \textstyle\sum_{m \in \mathcal{M}} B_{\text{VLC},m}) \right]^+;$$

end for

Output: $B_{\text{VLC},m} \forall m \in \mathcal{M}$

Similarly, given the allocated RF power from Algorithm 6.3.12, the optimal RF bandwidth can be allocated using Algorithm 6.3.14.

Algorithm 6.3.14 Bandwidth Allocation for RF network

Input: $P_{\text{RF},m}$, μ_m, and λ;

Initialization: $\beta_{\text{RF}}(1) \geq 0$;

for $i = 1 : I$ **do**

for $m \in \mathcal{M}$ **do**

Find $B_{\text{RF},m}(i)$ as the positive real root of $\kappa_{\text{RF}} \left(\log_2 \left(1 + \frac{h_{\text{RF},m}^{\text{LoS}} P_{\text{RF},m}}{B_{\text{RF},m} N_{0,\text{RF}}} \right) - \right.$

$$\left. \frac{P_{\text{RF},m} h_{\text{RF},m}^{\text{LoS}}}{\ln(2) \left(B_{\text{RF},m} N_{0,\text{RF}} + P_{\text{RF},m} h_{\text{RF},m}^{\text{LoS}} \right)} \right) + (1 - \kappa_{\text{RF}}) \left(\log_2 \left(1 + \frac{h_{\text{RF},m}^{\text{NLoS}} P_{\text{RF},m}}{B_{\text{RF},m} N_{0,\text{RF}}} \right) \right.$$

$$\left. - \frac{P_{\text{RF},m} h_{\text{RF},m}^{\text{NLoS}}}{\ln(2) \left(B_{\text{RF},m} N_{0,\text{RF}} + P_{\text{RF},m} h_{\text{RF},m}^{\text{NLoS}} \right)} \right) = \frac{\beta_{\text{RF}}}{1 + \mu_m};$$

end for

$$\beta_{\text{RF}}(i+1) = \left[\beta_{\text{RF}}(i) - \eta_4 (B_{\text{RF,max}} - \textstyle\sum_{m \in \mathcal{M}} B_{\text{RF},m}) \right]^+;$$

end for

Output: $B_{\text{RF},m} \forall m \in \mathcal{M}$

Using the optimal power and bandwidth allocated in Algorithms 6.3.11–6.3.14, the objective now is to jointly allocate the resources that satisfy the target data rate and maximize the resulting energy efficiency for a given λ. Algorithm 6.3.15 gives the optimal joint solution of (6.22) for a given value of λ. In Algorithm 6.3.15, we iterate over the power and bandwidth allocations until convergence to find the optimal joint bandwidth and power allocation solution that maximizes the energy efficiency in the heterogeneous VLC–RF network and satisfies the required QoS by all MTs. The optimal solution accounts for the RF and VLC characteristics (reliability and energy efficiency metrics).

Algorithm 6.3.15 Joint Power and Bandwidth Allocation for the RF–VLC Network

---

Input: λ;

Initialization: $\mu_m(1) \geq 0$;

for $i = 1 : I$ **do**

Allocate power of the VLC network using Algorithm 6.3.11;

Allocate power of the RF network using Algorithm 6.3.12;

Allocate bandwidth of the VLC networkusing Algorithm 6.3.13;

Allocate bandwidth of the RF network using Algorithm 6.3.14;

for $m \in \mathcal{M}$ **do**
$$\mu_m(i+1) = \left[\mu_m(i) - \delta_5(R_{\mathrm{VLC},m} + R_{\mathrm{RF},m} - R_{\min,m})\right]^+;$$
end for
end for
Output: $P_{\mathrm{VLC},m}, P_{\mathrm{RF},m}, B_{\mathrm{VLC},m}, B_{\mathrm{RF},m} \forall m \in \mathcal{M}$

Denote by I_r the number of required iterations for the Algorithm 6.3.10 to converge. The complexity of the proposed resource allocation strategy is determined by calculating the number of dual variables in the dual problem. Therefore, the resource allocation framework computational complexity is given by $O(I_r M)$, which is linear in the number of the MTs. The number of MTs that could be accessed using a single VLC AP for an indoor environment is small, and therefore the computational complexity of the framework is reasonably low.

6.3.2   Performance Evaluation

Next, the energy efficiency of the integrated RF–VLC heterogeneous network (which is referred to in the simulation results by 'RF–VLC') is compared with two benchmark systems. The first benchmark represents a system consisting of a single RF wireless network (denoted by 'RF-Only') and hence no multi-homing is performed. The second benchmark represents a system comprising two RF APs over different frequency bands (denoted by 'RF–RF') and hence multi-homing is achieved only over RF links. In the system with two RF APs, one of the RF systems is assigned a bandwidth equal to that of the VLC system to ensure a fair comparison. In the following simulation results, we assume $R_m = 2\,\mathrm{Mbps}$, $P_{\mathrm{RF}} = 1\,\mathrm{W}$, $P_{\mathrm{RF}}^c = 6.7\,\mathrm{W}$, $P_{\mathrm{VLC}}^c = 4\,\mathrm{W}$, $N_{0,\mathrm{rf}} = 3.89 \times 10^{-21}\,\mathrm{W/Hz}$, $N_{0,\mathrm{VLC}} = 10^{-21}\,\mathrm{W/Hz}$, $B_{\mathrm{RF}} = 10\,\mathrm{MHz}$, $B_{\mathrm{VLC}} = 20\,\mathrm{MHz}$, $\varsigma = 10\,\mathrm{W/amp}$, $\chi = 0.8\,\mathrm{amp/W}$ and $M = 4$. MTs are uniformly distributed away from the RF AP in the range between 1 and 1.5 m and are uniformly distributed away from the VLC AP in the range between 1.5 and 2 m. The maximum power of the VLC AP is given by the product of the number of LEDs used at the VLC source with the maximum power driving each LED. The number of LEDs is set to 38 with the maximum power to drive a LED to 300 mW. This value generates around 900 lumens from the VLC source, which is practically a suitable value for lighting.

Figure 6.4 shows the energy efficiency of the different systems against the number of MTs. The performance of the RF–VLC system is significantly better than the performance of the RF-only system thanks to the multi-homing capability of the MTs and the energy-efficient communication nature of the VLC AP. Through the multi-homing capability, MTs can enjoy communication links with better channel conditions with at least one AP, leading to high data rates and low power consumption. In addition, the performance of the RF–VLC system is better than that of the RF–RF system due to the lower power consumption cost of the VLC AP than the RF AP. The power consumption in RF APs is the sum of the fixed and transmission powers, while in the VLC AP, the power is due only to the fixed power component, since no power is dedicated for transmission as its transmission power is already used for illumination.

Figure 6.5 shows a plot of energy efficiency performance against the fixed power consumption of the VLC AP to investigate the effect of any increased fixed power consumption in the VLC AP. The energy efficiency of the RF–VLC system is equal to that of the RF–RF system when the fixed power is 6 W, which is nearly equal to the fixed power of an RF AP.

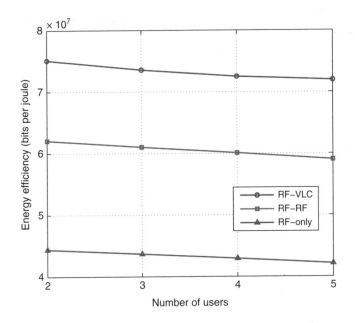

Figure 6.4   Energy efficiency versus the number of MTs [183]

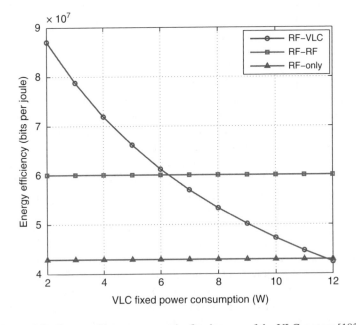

Figure 6.5   Energy efficiency versus the fixed power of the VLC system [183]

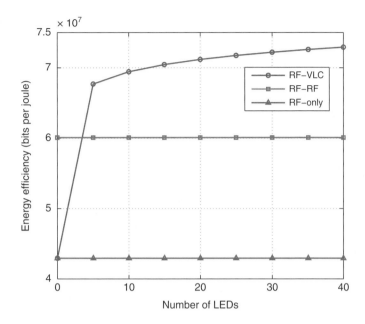

Figure 6.6   Energy efficiency versus the number of LEDs used by the VLC system [183]

One can conclude that the integration of a VLC AP in a heterogeneous networking with RF APs is beneficial only if the VLC AP fixed power consumption is less than an equivalent RF AP.

Figure 6.6 investigates the effect of the number of LEDs on the energy efficiency of the RF–VLC system. Increasing the number of LEDs allows a higher transmission power for the VLC system, which motivates the MTs to obtain most of their required data service from the VLC AP, reducing the transmission power consumption of RF AP, and hence improving the overall energy efficiency. In addition, the figure shows that introducing the VLC network even with a small number of LEDs (only 5 LEDs) enhances the energy efficiency significantly.

Figure 6.7 studies the case in which the LoS availability probabilities for RF and VLC systems are equal and shows the energy efficiency versus the LoS availability probability. The performance of the RF–VLC system is better than the benchmarks when the probability of LoS availability in the VLC AP is higher than 0.7 because of the good energy efficiency properties of the proposed RF–VLC system. Furthermore, the slope of the curve of the RF–VLC energy efficiency is higher than those of the benchmark systems because of the significance of the LoS availability in the VLC system compared with the RF system.

Figure 6.8 investigates the effect of LoS availability probability in the RF system on the energy efficiency of the RF–VLC heterogeneous system when $\kappa_{\mathrm{VLC}} = 1$. In the RF–VLC system, the MTs exploit the less costly VLC energy for data transmission and exploit the RF transmission power when needed. Consequently, the performance improvement with the increase of RF LoS availability probability presents a smaller slope in the integrated system than the RF benchmarks.

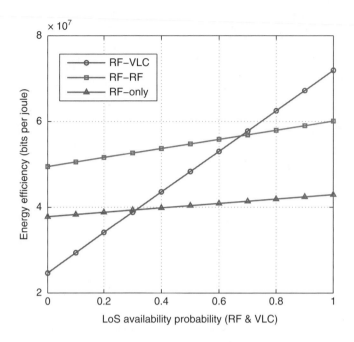

Figure 6.7   Energy efficiency versus the LoS availability probability in VLC and RF systems [183]

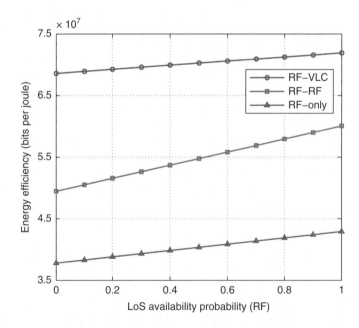

Figure 6.8   Energy efficiency versus the LoS availability probability in RF systems [183]

6.3.3   Green VLC–RF Internetworking Challenging Issues

There are a number of challenging issues that should be addressed to enable more efficient integration of VLC and RF APs in a heterogeneous wireless network. In order to overcome these challenging issues, further investigations are required to deal with the users' spatial distribution, the APs' placement and the interference concerns in VLC networks. These issues are discussed next in more details.

6.3.3.1   MT Spatial Distribution and AP Placement

High data rates in VLC APs can be achieved when there is an unobstructed LoS between transmitter and receiver. However, the data rate is reduced significantly in the absence of LoS, which can be due to the MT FoV misalignment with the VLC AP or due to obstacles. Consequently, data rate can be abruptly degraded once the LoS is obstructed, unlike the RF networks. Hence, in VLC APs, the network performance is very sensitive to the probability of the LoS availability, which in turn depends on the specific MT spatial distribution. The probability of the LoS availability can vary significantly with any slight change in the user locations. Therefore, the exact knowledge of the spatial distribution of the MTs is a crucial parameter in assessing the performance of the VLC network. Thus, the performance of the resource allocation mechanism in heterogeneous networks integrating RF and VLC APs is highly affected by the spatial distribution of the MTs. Hence, allocating network resources in the RF and VLC networks given the knowledge of the MTs spatial distribution can lead to better energy efficiency. Therefore, studying the MTs' spatial distribution and the techniques to employ this knowledge in resource allocation problems plays an important role in the system design stage.

Furthermore, it is necessary to design a resource allocation framework that is robust to uncertainties in MTs' locations. For instance, it is shown in [193] that a small error in the user location information can degrade significantly the data rate obtained from the VLC AP. Moreover, the achieved data rate at the VLC communication dead zones is relatively low. By examining the spatial distribution of MTs and studying the expected required rates of the users located in the VLC communication dead zones, more RF transmission power can be allocated to the MTs in these VLC dead zones while allocating more VLC transmission power to the VLC covered regions, which calls for a joint design of the VLC and RF APs coverage areas. The selection of the positions of the VLC and RF APs in both systems can considerably affect the system performance. While the RF AP placement is constrained mainly by the communication requirements, the VLC AP position is constrained by both the illumination and communication requirements [194]. Moreover, the transmission powers in both VLC and RF communication networks are highly affected by the path loss, and hence finding the optimal placement of the APs can highly benefit the achieved energy efficiency. Further research is needed for the joint placement optimization of RF and VLC APs to improve the energy efficiency while maintaining the communication and illumination requirements.

6.3.3.2   Interference Issues in VLC Networks

Communication over the light spectrum can be exposed to different types of interferences due usage of various light sources, for example, the ambient light. The visible light spectrum is wide enough to allow high data rates. In indoor scenarios, light does not penetrate walls.

Hence, any portion of the visible light spectrum can be exploited inside some closed indoor spaces. Being unlicensed allows exploiting the spectrum for various services using different communication schemes. On the contrary, being unlicensed does not allow reserving certain portions of the spectrum for communication purposes only, and hence any light source or light reflection is considered as an interference source in VLC networks. This problem can be addressed by employing coding approaches in VLC systems. However, such an approach affects the energy consumption in VLC systems due to the extra processing requirements. By integrating both VLC and RF systems, interference mitigation solutions that exploit data transmissions over the two used spectra can improve energy efficiency. However, such a solution requires a further investigation.

6.4 Summary

Integrating RF and VLC APs into a heterogeneous wireless networking environment is a promising solution to support energy-efficient communications. The multi-homing capability of the MTs in a heterogeneous network with VLC and RF APs enables users to benefit from the huge unlicensed bandwidth of the visible light spectrum and the low cost of power transmission. Also, it provides a reliable means of communication in VLC systems, as RF communications are employed in the absence of VLC LoS. In this chapter, the superior performance of the heterogeneous integrated RF–VLC network is demonstrated and compared with the benchmarks that support an RF-only network or a heterogeneous network consisting of two RF systems, respectively. Further investigations are required to deal with the joint RF–VLC planning design (AP placement), exploit the MT spatial distribution information in the radio resource allocation and cope with the interferences present in a heterogeneous network with multiple VLC APs to support even greener communications.

Part Three

Network Management Solutions

7

Dynamic Planning in Green Networks

This chapter investigates operation of green wireless networks by using a dynamic planning approach. At a low-call traffic load, network operators are expected to save energy by switching off some of their BSs, and letting mobile users to be served by the remaining active BSs. In this context, three research directions are discussed. The first direction examines dynamic cell zooming and BS sleep scheduling for dense heterogeneous networks with macro and pico BSs. The second direction studies cooperation of network operators as a means of energy saving, where networks with overlapped coverage cooperate to reduce their energy consumption by alternately switching on and off their resources according to the traffic load conditions. Finally, the third direction presents a dynamic planning framework with balanced energy efficiency that accounts for the energy consumption of the mobile users in the uplink in addition to that of the network operators in the downlink.

7.1 Introduction

Great advancements in wireless communications services have resulted in high energy consumption by network operators and mobile users. In the literature, there have been several proposals for designing an energy aware infrastructure in wireless communications networks. Energy awareness in wireless communications networks has been studied for a long time in mobile devices and wireless sensors, due to their limited power capabilities. Recently, such awareness has been extended to cellular network BSs due to financial and environmental considerations. In this regard, researchers have proposed to exploit the traffic load (temporal and spatial) fluctuations, by switching off some of the available radio resources when the traffic load is light. Such an approach is known as dynamic planning. The investigated resources can be the radio transceivers of active BSs [195]. When a BS is in its active mode, power supply, processing circuits and air conditioning consume up to 60% of the total energy [34]. Therefore, significant energy saving can be achieved if the entire BS is switched off when the traffic load is light [34]. In this chapter, three research directions will be investigated,

Green Heterogeneous Wireless Networks, First Edition. Muhammad Ismail, Muhammad Zeeshan Shakir, Khalid A. Qaraqe and Erchin Serpedin.
© 2016 John Wiley & Sons, Ltd. Published 2016 by John Wiley & Sons, Ltd.

namely dynamic planning with dense small-cells, network cooperative dynamic planning and balanced dynamic planning. These approaches will be the major focus of this chapter.

A heterogeneous wireless network includes a mix of various cell sizes and shapes, such as high-power macro cells and low-power nodes such as pico cells and relays. It is expected that a dense deployment of low-power BSs (small-cells) will be implemented in the near future. However, the dense deployment of low-power BSs raises several fundamental issues in terms of energy consumption. Particularly, key issues for energy efficient deployment of dense small-cells include finding the optimal cell-zooming techniques (i.e. expanding and shrinking the cell size) and sleep policies (i.e. on–off switching of BSs) for both macro and small (pico) BSs depending on the traffic pattern. In this chapter, we present optimal energy-efficient and QoS-aware dynamic cell-zooming and BS-switching policies for dense small-cell deployment subject to constraints on user data rate requirements and outage probability [196].

Most of the dynamic planning solutions are limited to the operation of a single network (as described in the previous paragraph). Consequently, the proposed solutions for service provision for the off cells rely on the active resources of such a network. This in turn leads to an increase in the transmission power, and hence an increase in inter-cell interference, and may result in coverage holes. Furthermore, most of the dynamic planning solutions in the literature focus on switching off either some of the BSs or some of the radio resources of the BSs. In this chapter, we present an optimal resource (BSs and radio channels) on–off switching framework that enables network operators with BSs of overlapped coverage in a given geographical region to cooperate among each other to achieve energy saving [12].

Finally, the existing research focuses only on improving energy efficiency for network operators and does not account for the incurred energy consumption for the mobile users in the uplink. In practice, dynamic planning can lead to higher energy consumption for mobile users in the uplink due to switching off nearby BSs and hence MTs suffer from larger transmission distances. As a result, MTs suffer from battery drain at a higher rate and hence are subject to call droppings. In this chapter, we present a dynamic planning approach that can capture and balance the trade-off in energy efficiency between network operators (in the downlink) and mobile users (in the uplink) [197, 198].

7.2 Dynamic Planning with Dense Small-Cell Deployment

Several works have addressed energy-efficient cell-zooming mechanisms and sleep policies for heterogeneous networks with dense small-cell deployment. In particular, energy minimization in macro-relay networks has been studied in [199] and [200]. However, both these contributions are restricted to a single macro cell scenario. The energy consumptions of a two-hop relaying scheme (LTE type 1 relay) and a multicast cooperative scheme (LTE type 2 relay) are studied in [201]. In spite of the interesting results presented in the above studies, the energy-efficient cell-zooming and BS-switching policies in the previous works do not guarantee the minimum data rate requirements and outage probability of end user. Moreover, these techniques cannot be generalized to dense small-cell deployment scenarios. We present a dynamic planning approach to optimally associate users to access nodes (pico and macro BSs) by adjusting the coverage area of these BSs. The objective of the optimal cell zooming and user association is to reduce the traffic load at some BSs and let them enter into sleep mode, while ensuring the user data rate demand.

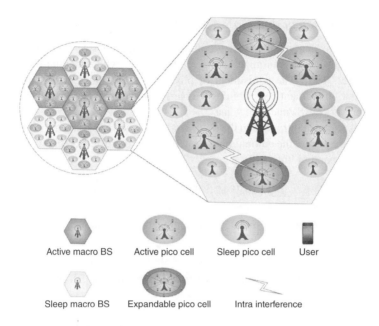

Figure 7.1   Dense macro–pico network [196]

Consider a downlink dense macro–pico network consisting of macro BSs, low-power pico BSs and MTs. An overview of a dense macro–pico network is shown in Figure 7.1. Denote the sets of macro BSs, pico BSs and MTs by $\mathcal{S}_m$, $\mathcal{S}_p$ and $\mathcal{M}$, respectively. Let x_{sm} be a binary variable representing the association between BS s and MT m. The achieved data rate in the downlink for MT m is given by

$$R_{sm} = x_{sm} B_{sm} \log_2(1 + \gamma_m), \quad \forall s \in \mathcal{S}_m \cup \mathcal{S}_p, m \in \mathcal{M}, \tag{7.1}$$

where γ_m denotes the received SINR at the MT and B_{sm} is a real variable representing the allocated bandwidth to MT m from BS s. The total bandwidth available at BS s is B_s. Let R_m represent user m required data rate. The total load at BS s is given by

$$\rho_s = \sum_{m \in \mathcal{M}} x_{sm} \frac{B_{sm}}{B_s}, \quad 0 < \rho_s \leq 1. \tag{7.2}$$

The BS power consumption model is given by

$$P_s = P_{f,s} + \xi_s P_s^{\text{out}}, \quad 0 < P_s^{\text{out}} \leq P_s^{\text{max}}, \tag{7.3}$$

where $P_{f,s}$ is the minimum power consumed when the BS is in idle mode and depends on the transceiver electronics, cooling and so on; ξ_s denotes the power amplifier efficiency of BS s and P_s^{out} is the maximum RF output power of BS s at full load and $P_s^{\text{out}} = \rho_s P_s^{\text{max}}$. Let y_s denote a binary variable representing BS s operation mode, where $y_s = 1$ represents an active BS and $y_s = 0$ denotes an inactive BS. Furthermore, $z_s \in [0, 1]$ is a real variable representing the cell-zooming condition of BS s by dynamically controlling BS s transmission power based on the power requirements of the farthest user in the cell.

7.2.1   Energy-Efficient and QoS-Aware Cell Zooming

Next, we present an analytical model for optimizing the cell size in a dense macro–pico heterogeneous network to minimize energy consumption while maintaining the end user quality of service [196]. The adjustment of physical cell parameters such as transmission power can help to implement cell zooming. In particular, cells can zoom in by decreasing the transmission power of BSs and vice versa.

The first constraint we introduce in the optimization framework concerns capacity and radio resource allocation, where the amount of allocated bandwidth by each BS s cannot exceed the BS total available bandwidth, that is,

$$\sum_{m \in \mathcal{M}} x_{sm} B_{sm} \leq B_s, \quad \forall s \in \mathcal{S}_m \cup \mathcal{S}_p. \tag{7.4}$$

Furthermore, no MT can be associated with an inactive BS, that is,

$$x_{sm} \leq y_s, \quad \forall s \in \mathcal{S}_m \cup \mathcal{S}_p, m \in \mathcal{M}. \tag{7.5}$$

In addition, each active user should be associated with only one BS at a time, that is,

$$\sum_{s \in \mathcal{S}_m \cup \mathcal{S}_p} x_{sm} = 1. \tag{7.6}$$

The allocated data rate to MT m should satisfy the user-required data rate, that is,

$$\sum_{s \in \mathcal{S}_m \cup \mathcal{S}_p} x_{sm} R_{sm} \geq R_m, \quad \forall m \in \mathcal{M}. \tag{7.7}$$

In order to serve cell edge MTs (at the maximum distance $d_{\max}$), the cell-zooming mechanism should satisfy

$$z_s P_s^{\text{out}} \geq \frac{P_r}{\kappa} d_{\max}^\alpha, \tag{7.8}$$

where P_r represents the received power required for the farthest user at distance $d_{\max}$, κ denotes a proportionality constant and α denotes the path loss exponent.

In the dynamic cell-zooming mechanism, the transmission power is adjusted according to a desired cell radius; therefore, a user may experience outage when it is out of a BS coverage area due to its movement during the cell-zooming operation. Formally, the outage probability is defined as

$$P_{\text{outage}} = \Pr(\gamma_m \leq \zeta), \tag{7.9}$$

where ζ denotes the SINR threshold.

The optimization framework objective is to minimize the total power consumed by the network while satisfying the users' target QoS, and it is expressed as [196]

$$\min_{x_{sm}, y_s, z_s, B_{sm}} \sum_{s \in \mathcal{S}_m \cup \mathcal{S}_p} y_s P_{f,s} + \xi_s z_s P_s^{\text{out}} \tag{7.10}$$

$$\text{s.t.} \quad (7.4)–(7.9).$$

The energy-efficient dynamic cell-zooming linear program (LP) problem in (7.10) can be solved optimally using the state-of-the-art CPLEX mixed integer program (MIP) solver. However, large instances of the above problem are difficult to be solved optimally. It turns out from the objective function and constraints in (7.10) that the total number of terms inside the constraints can increase significantly with the increase in the number of macro BSs, pico BSs and users due to the fact that the number of connection opportunities among different nodes increases with a growth in the density of nodes. Therefore, the computational complexity for finding the CPLEX MIP optimal solution increases dramatically as larger clusters of cells are considered. The increased computational complexity issue requires designing approximate solution strategies. In this regard, a practical and effective approach is to apply a distributed technique for finding the solution, in which repetitive uses of small-cell clusters can be made to deal with a larger cluster [196]. For small-cell clusters, the problem can be solved efficiently in a distributed manner. Then, one can merge the distributed solutions to obtain the network-wide solution. Clearly, a dramatic reduction in computational complexity comes at the cost of a performance gap between the optimal CPLEX solution and the distributed solution.

7.2.2 Performance Evaluation

In this section, we evaluate the performance of the energy-efficient cell-zooming mechanism. The parameter values used in the analysis are reported in Table 7.1. The first scenario consists of a seven-cell macro–pico network with hotspot and uniformly distributed users.

Figure 7.2 depicts the effect of the density of pico BSs on the number of active macro base stations in operation. From the plot depicted in Figure 7.2, it can be seen that the dynamic cell-zooming algorithm for a dense macro–pico network can offload the traffic from macro BSs by increasing the coverage area of pico BSs and switching off most of the lightly loaded macro base stations. Moreover, it is evident from Figure 7.2 that more macro BSs are switched off by increasing the density of pico BSs in the network.

Table 7.1 System parameters [196]

Parameter	Value
Carrier frequency	2 GHz
Bandwidth	20 MHz
Thermal noise PSD	-174 dBm/Hz
Carrier frequency	2 GHz
$P_s^{\max}, s \in \mathcal{S}_m$	40 W
$P_{f,s}, s \in \mathcal{S}_m$	712 W
$\xi_s, s \in \mathcal{S}_m$	14.5
$P_s^{\max}, s \in \mathcal{S}_p$	1 W
$P_{f,s}, s \in \mathcal{S}_p$	14.9 W
$\xi_s, s \in \mathcal{S}_p$	8
MT transmission power	23 dBm
Antenna configuration	TX-1, RX-1
$\mathrm{PL}_{\mathrm{LoS}}(d)$: Macro-MT link	$103.4 + 24.2 \log_{10}(d)$
$\mathrm{PL}_{\mathrm{NLoS}}(d)$: Macro-MT link	$131.1 + 42.8 \log_{10}(d)$
$\mathrm{PL}_{\mathrm{LoS}}(d)$: Pico-MT link	$103.8 + 20.9 \log_{10}(d)$
$\mathrm{PL}_{\mathrm{NLoS}}(d)$: Pico-MT link	$45.4 + 37.5 \log_{10}(d)$
Inter-site distance	500 m

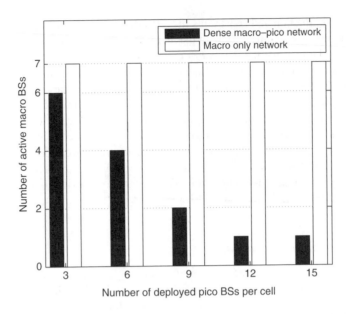

Figure 7.2   Number of active macro BSs for a dense macro–pico network [196]

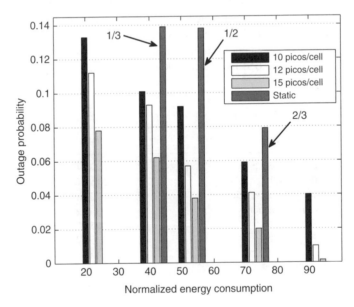

Figure 7.3   Trade-off between outage probability and energy consumption for a dense macro–pico network [196]

Figure 7.3 depicts that there is a trade-off between outage probability and energy consumption. It must be mentioned here that in this plot the energy consumption is normalized to 100 if all the BSs are in active mode. It is evident from Figure 7.3 that the outage probability decreases by increasing the density of pico BSs. However, this decrease in outage probability

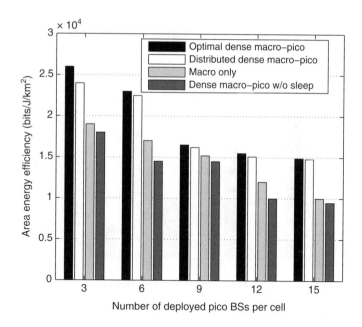

Figure 7.4 Area energy efficiency for a dense macro–pico network [196]

comes at the cost of slightly increased energy consumption. We also compare the performance of the dynamic cell-zooming algorithm with a static cell-zooming algorithm, which reduces the coverage area of BSs to $1/2$, $1/3$ or $2/3$ of the maximum coverage. Figure 7.3 shows that the dynamic cell-zooming algorithm performs better than the static algorithm. Moreover, the dynamic cell-zooming algorithm is more flexible as it can freely leverage the trade-off between outage probability and energy consumption.

A plot of the area energy efficiency against the total offered traffic load for different dense small-cell scenarios is depicted in Figure 7.4. The bar labelled *optimal dense macro–pico network* in Figure 7.4 represents the scenario where we take into account transmission and circuit energy of macro and pico base stations in active mode, which are obtained by solving the problem in (7.10) using CPLEX. Moreover, the bar labelled *distributed dense macro–pico network* represents the scenario where we solve the problem in (7.10) using a cluster based distributed approach. The bar labelled *dense macro–pico network without sleep mode* represents the scenario where we consider transmission and circuit energy of all macro base stations in the network together with the deployed pico cells. Finally, the bar *macro only* refers to the scenario where we consider only macro BSs, and all macro BSs are assumed in the active mode. It is evident from Figure 7.4 that the macro–pico network has the best area energy efficiency compared with other cases. Figure 7.4 illustrates that the area energy efficiency of the optimal dense macro–pico solution is significantly higher than that corresponding to the macro-only case. The rationale behind this fact is that most of the macro BSs are operating in an inactive mode in the optimal dense macro–pico network, as shown in Figure 7.4, as compared with other small-cell scenarios.

The user association for a seven-cell dense macro–pico network powered by a stand-alone power grid is depicted in Figure 7.5. It turns out that the dynamic cell-zooming algorithm

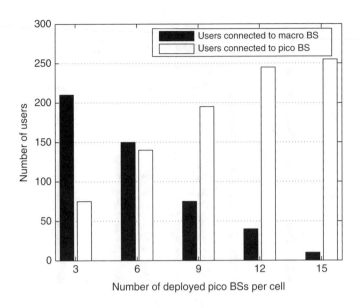

Figure 7.5   User association for a dense macro–pico network [196]

adjusts the transmission power from the macro and pico BSs in such a way that more users connect to the pico BSs than to the macro BSs with an increase in the density of pico BSs in the network.

7.3   Dynamic Planning with Cooperative Networking

While BS on–off switching can avoid resource over-provision in a low-traffic load condition and hence achieve energy saving, the radio coverage and service provision for the off cells face several challenges. Since most of the dynamic planning solutions are limited to the operation of a single network, the proposed solutions for service provision for the off cells rely on the active resources of such a network, as mentioned in the previous section. Consequently, an increase in the transmission power of the active BSs is required to increase the radii of its cells to provide radio coverage for the off cells. This may also result in coverage holes if the maximum allowed transmission power of the remaining active BSs cannot achieve radio coverage for the switched-off cells, and therefore, service disruption is expected in these areas. Also, an increase in the transmission power may result in inter-cell interference in the case that more than one active BS tries to achieve radio coverage for the switched-off cells. Consequently, additional interference management schemes are needed. With the existence of different network operators with overlapped coverage among different BSs, network cooperation can achieve energy saving and avoid the dynamic planning shortcomings. In addition, the dynamic planning solutions proposed in the literature focus on either switching off some of the BSs or switching off some of the radio resources of the BSs. It is more beneficial to combine both strategies and not only switch off some BSs but also switch off some of the radio resources of an active BS to further improve the amount of energy saving. In this case, an optimal resource (BSs and radio channels) on–off switching framework

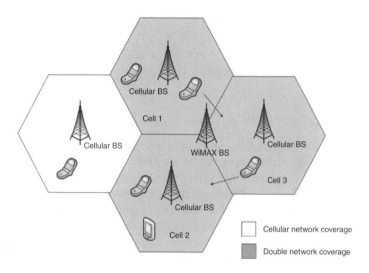

Figure 7.6   Network coverage areas [12]

is required to determine the operation mode of BSs and radio channels that adapts to the fluctuations of the traffic load and maximizes the amount of energy saving under service quality constraints, assuming a cooperative networking environment.

Consider a geographical region with two network operators [12]. The first network operator offers 4G cellular services while the second operator offers WiMAX services. The cellular network covers the whole geographical region. In regions with a high service demand, WiMAX BSs are deployed to provide high capacity. Let S denote the set of cellular network cells covered by the WiMAX network BS, $S = \{1, 2, \ldots, S\}$ as shown in Figure 7.6. An MT in the overlapped coverage can be served by either of the two networks, that is, no multi-homing is allowed. Let C_c denote the number of channels available in a cellular network BS. The WiMAX network BS has C_w channels. Each channel has a fixed bandwidth B. For simplicity of illustration, assume that a call requires one channel from one of the networks for its service. A network BS working mode variable x_s for the cellular network BS ($s \in S$) is represented by a binary digit "0" to indicate an inactive (off) BS or "1" to indicate an active (on) BS. Similarly, the binary variable x_{S+1} is used for the WiMAX network BS to represent its on/off states. There is always at least one active network in the overlapped coverage area to guarantee service provision. Let $X = [x_1 x_2 \cdots x_{S+1}]$ denote a vector of BS working modes in the overlapped coverage area. The total power consumption of a BS is denoted by P_w (P_c) for WiMAX (cellular) network, and consists of two components. A fixed component which accounts for the BS power supply and air conditioning is represented by $P_{f,w}$ ($P_{f,c}$) for a WiMAX (cellular) network. The variable component depends on the number of active channels in the BS, and accounts for the power amplifier, feeder loss and transmitted power, and is represented by $P_{v,w}$ ($P_{v,c}$) for WiMAX (cellular) network. The number of active channels in cell s is given by $k_{s,w}$ ($k_{s,c}$) for WiMAX (cellular) network, $s \in S$. The power consumption of an inactive/sleep WiMAX (cellular) network BS is denoted by $P_{0,w}$ ($P_{0,c}$). When a BS changes its working mode from inactive to active, more energy is required to start up the BS power supply, circuits, and air conditioning. The switching cost is represented by an additional power consumption β of the BS fixed-power component.

The following traffic and mobility assumptions are made: (A1) The aggregate traffic arrivals (new and handoff calls) to the cell are modelled by a Poisson process with mean arrival rate λ_s; (A2) The channel-holding time in the cell (the minimum of the user cell dwell time and the call duration) follows an exponential distribution with mean $1/\mu_s$.

7.3.1 Optimal Resource On–Off Switching Framework

Network cooperation in green radio communications exploits the temporal fluctuations in the traffic load to save energy. This is achieved by alternately switching on and off the available resources from BSs of different networks in regions with overlapped coverage, according to the traffic load condition. In general, two types of traffic load fluctuations can be distinguished: (i) large-scale fluctuation, in which the traffic load varies significantly from one period to another along the day and (ii) small-scale fluctuation, in which the traffic load varies slightly around some average value. Hence, we partition the time into a set of periods $\mathcal{T} = \{1, 2, \ldots, T\}$ of constant duration τ (h) that can capture the large-scale fluctuations in the traffic load along the day, $T = \lceil 24/\tau \rceil$. Each period $t \in \mathcal{T}$ is further partitioned into a set $\mathcal{I} = \{1, 2, \ldots, I\}$ of smaller periods each of duration Λ to capture the small-scale fluctuations of the traffic load during that period, $I = \lceil \tau/\Lambda \rceil$. The large-scale fluctuations in the traffic load can be exploited to turn off some BSs in a light-load condition and transfer the traffic load to the remaining active networks to save energy. On the contrary, the small-scale fluctuations can be exploited to switch off some of the radio channels in each active BS to further reduce the amount of energy consumption.

Decisions on the BS working mode are made at the initial moment of each period t. While BS on–off switching can save energy, a switching action that is not compatible with the traffic load during a given period will result in a high call-blocking probability. An appropriate switching decision should maximize the amount of saved energy during that period and, at the same time, should achieve acceptable service quality, for example, as in terms of call-blocking probability. Also, it is desirable to minimize the frequency at which a BS changes its working mode from inactive to active to avoid the switching cost due to the additional energy consumption required for the BS start-up. The aggregate traffic arrival rate for each period is estimated using the data of traffic arrivals observed in previous days, as the traffic load, in general, follows a repeating pattern every day. Call arrivals follow a Poisson distribution and call departures follow an exponential distribution. The call-blocking probability is calculated using the Erlang B loss model. The optimal BS on–off switching decision for a given period t can be obtained by carrying out the following optimization problem [12]

$$\max_{C_s>0,J,X} \left\{ v \left[\sum_{s=1}^{S}(P_c - P_s) + (P_w - P_{S+1}) \right] - \right.$$

$$\left. (1-v) \left[\sum_{s=1}^{S} \Delta P_s + \Delta P_{S+1} \right] \right\}$$

$$\text{s.t.} \quad \frac{(\lambda_s/\mu_s)^{C_s}/C_s!}{\sum_{c_s=1}^{C_s}((\lambda_s/\mu_s)^{c_s}/c_s!)} \le \epsilon \;\; \forall s \in \mathcal{S} \tag{7.11}$$

$$x_{S+1} = \begin{cases} 1, & \text{if } \exists C_s > C_c, s \in \mathcal{S} \\ 0, & \text{otherwise} \end{cases}$$

$$\sum_{s=1}^{S} x_s = \begin{cases} S, & \text{if } x_{S+1} = 0 \\ J, & \text{if } x_{S+1} = 1, \sum_{s=1}^{S} C_s \leq C_w + J C_c. \end{cases}$$

The objective function in (7.11) represents the total power saving in the overlapped coverage area. The variables P_s and P_{S+1} denote the BS power consumption for the cellular and WiMAX network, respectively, and depend on the BS working mode. Hence, $P_s = P_c$ if $x_s = 1$, and $P_s = P_{0,c}$ otherwise. Similarly, $P_{S+1} = P_w$ if $x_{S+1} = 1$, and $P_{S+1} = P_{0,w}$ otherwise. The variables ΔP_s and ΔP_{S+1} denote the additional power consumption required for the BS to start up. Hence, $\Delta P_s = \beta P_{f,c}$ if the cellular network BS changes its working mode from inactive to active, and $\Delta P_s = 0$ otherwise. Similarly, $\Delta P_{S+1} = \beta P_{f,w}$ if the WiMAX BS changes its working mode from inactive to active, and $\Delta P_{S+1} = 0$ otherwise. From the objective function definition, there exists a trade-off between the amount of energy saving achieved by switching on or off different BSs and the switching cost due to the additional energy consumption required for a BS to start up when its working mode changes from inactive to active. The parameter υ is a weighting factor that models the relative importance between energy saving and the BS start-up switching cost. The variable C_s represents the required number of channels in cell $s \in \mathcal{S}$, $\sum_{s=1}^{S} C_s \leq C_w + S C_c$. The first constraint in (7.11) guarantees an acceptable service quality in terms of call-blocking probability not larger than a required upper bound ϵ, where the value of λ_s for a given t is the largest aggregate traffic arrival rate over $\mathcal{I}$ in that t. The WiMAX BS working mode rule is given in the second constraint in (7.11), while the number of required active BSs from the cellular network is given in the last constraint in (7.11), with $J = \{0, 1, \ldots S\}$. The BS operation rules are designed to satisfy the service demand in each cell for a given λ_s and ensure radio coverage in the overlapped area. Hence, (7.11) results in the optimal BS working mode for the WiMAX and cellular network in the geographical region, which maximizes the amount of energy saving during some period t, limits the frequency at which BSs change their working mode and provides a satisfactory call-blocking probability. A search algorithm can be used to solve (7.11). In this case, the values of C_s, which violate the service quality constraint in (7.11), are excluded from the search space. Different working mode vector X values can be composed from the feasible C_s values using the BS operation rules in (7.11). Hence, the working mode vector X, which maximizes the objective function value of (7.11), can be found. If the large-scale optimization problem results in more than one optimal BS working mode vector X, the working mode vector X is chosen from these optimal vectors such that the cells with the lowest traffic loads are switched off.

For each active BS, we can further exploit the small-scale fluctuations in the traffic load to find the optimal number of active channels that maximizes the percentage energy saving for the active BS and achieves an acceptable call-blocking probability. This is calculated at the beginning of each period $i \in \mathcal{I}$ using the following cost function [12]

$$\max_{C_s > 0} \{x_s[P_c - (P_{f,c} + k_{s,c} P_{v,c})] + x_{S+1}[P_w - (P_{f,w} + k_{s,w} P_{v,w})]\}, \quad \forall s \in \mathcal{S} \quad (7.12)$$

where x_s and x_{S+1} are obtained from the solution of (7.11). The optimization problem (7.12) is subject to the service quality constraint in (7.11), where λ_s is defined for each $i \in \mathcal{I}$. With

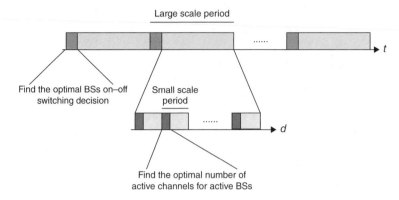

Figure 7.7 Time sequence of optimization events for the network cooperation energy-saving framework [12]

a larger coverage area of the WiMAX BS, it is assumed that the power consumption of each active radio channel in the WiMAX BS is not less than that of a cellular network BS (i.e. $P_{v,w} \geq P_{v,c}$). In this case, in order to further improve the amount of energy saving, when the BSs of both networks are active, more radio channels from the cellular network are utilized. As a result, we let $k_{s,c} = C_c$ and $k_{s,w} = C_s - C_c$ when BSs from both networks are active; otherwise, the active number of radio channels from the active BS is equal to C_s. The time sequence of optimization events is illustrated in Figure 7.7.

7.3.2   Performance Evaluation

In this section, we evaluate the performance of the cooperative networking dynamic planning approach using the framework depicted by (7.11) and (7.12). The geographical region under consideration is given in Figure 7.6 with the coverage of 3 cellular BSs that overlap with the coverage area of a WiMAX BS. It is assumed that the initial BS working mode vector is $X = 1111$. The system parameters are given in Table 7.2. The number of available radio channels in the cellular network and WiMAX BSs are chosen in a way that reflects the higher capacity of the WiMAX BS. The total number of available radio channels in the region are determined such that the peak traffic load does not violate the target level of the call-blocking probability. The different power components of the WiMAX BS are chosen such that they

Table 7.2 System parameters [12]

Parameter	Value	Parameter	Value	Parameter	Value
C_c	10	P_c	400 W	τ	1 h
C_w	72	$P_{f,c}$	250 W	Λ	15 min
P_w	1,500 W	$P_{0,c}$	10 W	υ	0.5
$P_{f,w}$	400 W	S	3	β	0.1
$P_{0,w}$	30 W	$1/\mu_s$	2.4 min	ϵ	0.01

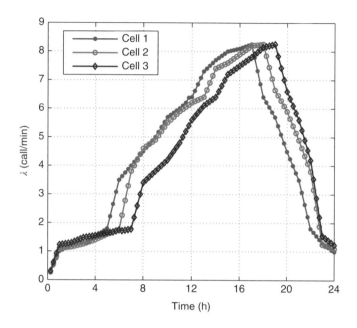

Figure 7.8 Aggregate traffic mean arrival rate in each cell [12]

Table 7.3 BS working mode [12]

Period	1–5	6–12	13–14	15–19	20	21–23	24
X	1110	0001	1001	1101	0101	0001	1110

are larger than that of a cellular BS, to reflect the fact that a WiMAX BS has more channels and covers a larger area than that of a cellular BS. The value of v gives equal importance for maximizing the amount of energy saving and reducing the BSs on–off switching cost.

Figure 7.8 shows the aggregate traffic mean arrival rate over the 24 h of a day for each cell. The λ values capture the traffic load fluctuations during the day. Variable λ exhibits a peak value in the middle of the day, while it assumes small values at early morning and late night periods.

The optimal decisions regarding the BS working mode for different periods are given in Table 7.3. The BS working mode varies in accordance with the traffic load fluctuations in each cell such that the optimal number of BSs is selected to maximize the amount of energy saving and yield a satisfactory quality of service.

The daily percentage of energy saving when all radio channels are active for the WiMAX BS is 24.5%, while for the cellular network BSs in cells 1, 2 and 3, the energy savings are 44.68%, 48.75% and 73.13%, respectively. With an optimized number of active channels, the daily percentage of energy saving for the WiMAX BS is 34.45%, and for the cellular network BSs in cells 1, 2 and 3 is 46.33%, 50.31% and 74.06%, respectively. This shows that the small-scale optimization problem significantly improves the amount of energy saving for the WiMAX BS.

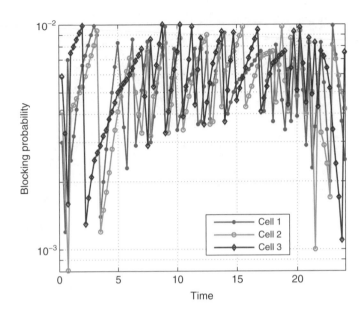

Figure 7.9   Call-blocking probability in each cell with the optimal number of active channels from the active BSs [12]

Figure 7.9 shows the call-blocking probability in each cell when the number of channels is optimized. The call-blocking probability in each cell has a desired maximum value $\epsilon = 10^{-2}$.

7.4   Balanced Dynamic Planning Approach

The main goal of dynamic planning is to reduce the BS energy consumption while ensuring an acceptable downlink service quality for mobile users. However, no attention is paid to the relation between the mobile users' perceived service quality and their incurred uplink energy consumption. Consequently, the BSs' switch-off decisions can result in energy-inefficient user associations from the mobile users' standpoint. As shown in Figure 7.10, accounting only for the downlink performance, MTs with uplink traffic can be associated to a faraway BS, due to a switched off nearby BS. Because of the long transmission distance, a high energy consumption in the MTs in the uplink is expected, which leads to energy depletion for MTs at a higher rate. Although energy consumption for MTs is not that much compared with the BS energy consumption, a high rate of battery depletion for MTs results in a higher rate of dropped services in the uplink, which jeopardizes the mobile users' perceived service quality. Dynamic planning approaches, if not carefully designed, can lead to higher energy consumption for the MTs in the uplink. In such a case, dynamic planning would only shift the energy consumption burden from the BSs to the MTs, which results in battery drain for MTs at a higher rate. Consequently, this will degrade the service quality perceived by the mobile users, for example, lower throughput, higher call dropping rate and so on. Hence, the future designs of dynamic planning should capture and balance the trade-off in energy efficiency among network operators and mobile users.

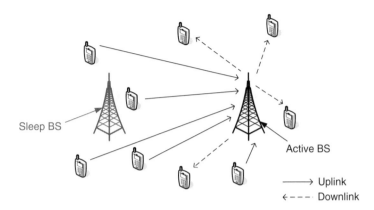

Figure 7.10 Dynamic planning with unbalanced energy saving [197]. MTs with uplink traffic are associated with faraway BSs

One challenge of dynamic planning is that the switching decisions of BSs are coupled with MT associations. Particularly, when a BS is switched off, the associated MTs will need to perform a handover process to another BS. Similarly, when a BS is turned on, the nearby MTs can perform a handover process to this BS. Also, newly incoming MTs are associated to a subset of active BSs to get service. However, the BS operation (i.e. on–off switching) does not occur at the same rate as the MT association. Hence, dynamic planning is a two-timescale problem. At a high level, the BS operation occurs at a low rate (with scale of hours) that depends on the call traffic load density. At a low level, the MT association takes place at a higher rate (with scale of minutes) based on user arrivals and departures. When only downlink traffic is considered, as in the existing research, the decisions at both levels are determined based only on the BS energy consumption, as shown in the previous two sections. With coexistence of uplink and downlink traffic, the decisions at both levels are determined based on the expected energy consumption at BSs and MTs.

When uplink traffic is considered, the BS switch-off decision criteria should be revised. Particularly, the switch-off decision criteria should capture the impact of MTs' battery drain on the uplink service degradation, for example, lower throughput, higher latency, higher call dropping rate and so on. Hence, the uplink service degradation is due to two factors, namely unavailability of radio resources at the BSs (due to BS switch-off) and MTs' battery drain (due to communicating with faraway BSs). The BS switch-off decision metric should balance the BS energy consumption with the uplink service quality due to MTs' battery drain. Similarly, the existing mechanisms employ the call traffic load increase as a wake-up decision criterion for a switched off BS [64]. However, in the presence of uplink traffic, the wake-up decision criteria should include, besides the call traffic load measure, a measure of MTs' service degradation due to battery drain. As a result, if the MTs' service quality is degraded due to battery drain, a nearby inactive BS should be turned on to avoid MTs' battery depletion and dropping of uplink calls.

Consider a geographical region that can be covered by a cluster S of BSs from different networks, as shown in Figure 7.11, $S = \{1, 2, \ldots, S\}$. The BSs of different networks operate in separate frequency bands, and there is no interference among them. Consider a cooperative networking scenario where different operators alternately switch on and off their

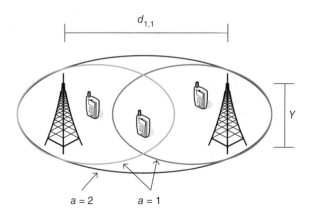

Figure 7.11 Example of dynamic planning cluster consisting of two BSs [197]. For simplicity, two tilting angles are assumed per BS leading to two coverage areas per BS

BSs to save energy while providing service coverage, as in the previous section. Let $d_{s,s'}$ denote the distance between BSs s and s', and define by Y_s the height of BS s, where $s, s' \in \mathcal{S}$. BSs control their coverage areas through antenna tilting [34]. For BS s, the tilting angle index is given by a_s, which corresponds to BS coverage area A_{s,a_s}. An inactive BS has $a_s = 0$.

Both uplink and downlink video call traffic loads are considered. Mobile users who capture live videos on their MTs and transmit them for online posting represent the uplink traffic [25]. On the contrary, mobile users performing video streaming represent the downlink video traffic. Let λ_{UL} and λ_{DL} denote the user arrival rates for uplink and downlink call traffic loads, respectively. The user arrival rates vary over the day and present peak values by mid-day and low values at early morning or late night periods [12]. Let μ_{UL} and μ_{DL} denote the average duration of the user service time in the uplink and downlink traffic, respectively. Each BS s can simultaneously support a maximum of M_s^{UL} and M_s^{DL} users in the uplink and downlink, respectively. Let R_{UL} and R_{DL} denote the minimum required data rate for uplink and downlink users, respectively. The BSs operate in frequency division duplex (FDD) mode, and variables B_s^{UL} and B_s^{DL} denote the available bandwidth for uplink and downlink users, respectively.

Let m_s^{DL} and m_s^{UL} denote the number of MTs served by BS s in the downlink and uplink, respectively, $m_s^{\mathrm{DL}} \leq M_s^{\mathrm{DL}}$ and $m_s^{\mathrm{UL}} \leq M_s^{\mathrm{UL}}$. Let $m^{\mathrm{DL}} = \sum_s m_s^{\mathrm{DL}}$ and $m^{\mathrm{UL}} = \sum_s m_s^{\mathrm{UL}}$ be the number of MTs in the geographical region with downlink and uplink traffic, respectively. The spatial distributions of MTs in the geographical region are described by the PMFs $\rho_{\mathrm{DL}}(A_{s,1})$ and $\rho_{\mathrm{UL}}(A_{s,1})$ for MTs with downlink and uplink traffic, respectively. The PMFs give the distribution of users in the proximity of each BS s, that is, in A_{s,a_s} with $a_s = 1$. For instance, $\rho_{\mathrm{DL}}(A_{1,1})m^{\mathrm{DL}}$ gives the number of MTs with downlink traffic in the proximity of the first BS.

Let Ω_{UL} and Ω_{DL} denote the average channel power gain in the uplink and downlink, respectively. The average channel power gain is characterized by the path loss model [202]

$$\mathrm{PL[dB]} = A \log_{10}(d[\mathrm{m}]) + B + C \log_{10}\left(\frac{f_c[\mathrm{GHz}]}{5}\right), \qquad (7.13)$$

where d represents the distance between the transmitter and receiver; f_c denotes the carrier frequency and A, B and C are model-dependent constants. The average path loss is determined based on the BS coverage area. Hence, for $a_s = 1$, as shown in Figure 7.11, d in (7.13) is dominated by Y_s, while for $a_s > 1$, d is dominated by $\sqrt{Y_s^2 + d_{s,s'}^2}$, where s' represents the BS that lies in BS s coverage, as shown in Figure 7.11. The average channel power gain is given by $10^{-\mathrm{PL[dB]}}/10$.

The average power consumption of BS s depends on its mode of operation, which is given by Auer et al. [39]

$$P_{s,\mathrm{dl}}(a_s, m_s^{\mathrm{DL}}) = \begin{cases} P_{F,s} + \Delta_s P_{s,\mathrm{tx}}(a_s, m_s^{\mathrm{DL}}), & a_s > 0 \\ P_{0,s}, & a_s = 0 \end{cases}, \qquad (7.14)$$

where $P_{F,s}$ is the BS fixed power consumption, which accounts for the power supply, cooling, backhaul and other circuits, Δ_s represents the slope of the load-dependent power consumption, $P_{s,\mathrm{tx}}(a_s, m_s^{\mathrm{DL}})$ denotes the BS average transmission power and $P_{0,s}$ denotes the BS sleep power. For BS s to support m_s^{DL} MTs in the downlink with minimum required data rate R_{DL}, its downlink transmission capacity should at least be $m_s^{\mathrm{DL}} R_{\mathrm{DL}}$. Using Shannon formula, the minimum BS average transmission power is expressed as

$$P_{s,\mathrm{tx}}(a_s, m_s^{\mathrm{DL}}) = \frac{N_0 B_s^{\mathrm{DL}}}{\Omega_{\mathrm{DL}}} \left(2^{\frac{m_s^{\mathrm{DL}} R_{\mathrm{DL}}}{B_s^{\mathrm{DL}}}} - 1 \right). \qquad (7.15)$$

In (7.15), $P_{s,\mathrm{tx}}(a_s, m_s^{\mathrm{DL}})$ is a function of a_s due to the effect of a_s on Ω_{DL}. The BS power consumption admits the maximum value $P_{s,\mathrm{mx}}$.

In order for BS s to support m_s^{UL} MT in the uplink with the minimum required data rate R_{UL}, the BS uplink transmission capacity should at least be $m_s^{\mathrm{UL}} R_{\mathrm{UL}}$. Using Shannon's formula, the minimum average power consumption of a given MT that is associated with BS s in the uplink is given by

$$P_{\mathrm{UL},s}(a_s, m_s^{\mathrm{UL}}) = P_c + \frac{N_0 B_s^{\mathrm{UL}}}{\Omega_{\mathrm{UL}}} \left(2^{\frac{m_s^{\mathrm{UL}} R_{\mathrm{UL}}}{B_s^{\mathrm{UL}}}} - 1 \right), \qquad (7.16)$$

where P_c denotes the MT circuit power consumption and the second part of (7.16) represents the average transmission power. The MT power consumption admits the maximum value of $P_{\mathrm{UL,mx}}$.

7.4.1 Two-Timescale Approach

In dynamic planning, the BS-switching decisions are coupled with the MT association decisions [35]. However, the BS operation and MT association decisions do not occur with the same rate. The BS operation decisions follow a low rate, with the scale expressed in hours, depending on the call traffic load density variation over the day. On the contrary, the MT association decisions occur at a high rate, with a scale expressed in minutes, depending on the user arrivals and departures. A two-timescale decision problem formulation can capture such a behaviour, as described in the previous section. Consequently, time is divided into slow and fast scales, as shown in Figure 7.12. At the slow timescale, time is partitioned into a set of periods $\mathcal{T} = \{1, 2, \ldots, T\}$ with fixed duration τ. Set $\mathcal{T}$ covers the 24 h of the day and captures the variations in the call arrival rates, that is, $\lambda_{\mathrm{UL}}(t)$ and $\lambda_{\mathrm{DL}}(t)$ (per unit τ) are fixed during period $t \in \mathcal{T}$ and vary from one period to another. Each period t is further partitioned into a set of periods $\mathcal{I} = \{1, 2, \ldots, I\}$ of equal duration Λ, $I = \lceil \tau/\Lambda \rceil$. $\mathcal{I}$ represents the fast

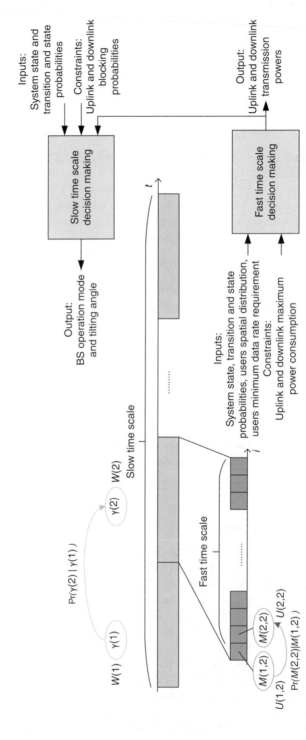

Figure 7.12 Illustration of the fast and slow timescales under consideration, the system states, actions, transition probabilities and the decision-making process [197]

timescale, and it captures the mobile users' arrivals and departures, that is, $m^{\mathrm{DL}}(i)$ and $m^{\mathrm{UL}}(i)$ are fixed during period $i \in \mathcal{I}$ and may vary from one period to another. The elements of the slow and fast timescale decision problems, namely states, actions, transition probabilities and cost will be explained next.

7.4.1.1   Slow Timescale

The slow timescale consists of a set of periods $\mathcal{T}$ that covers the 24 h of the day, each period having a long duration τ. The slow timescale decision problem is to control the BS operation (on and off) to minimize (and balance) the total expected energy consumption of network operators and mobile users.

The slow timescale system state $\Upsilon(t)$ yields the uplink and downlink call traffic load densities, where $\Upsilon(t) = (\gamma_{\mathrm{UL}}(t), \gamma_{\mathrm{DL}}(t))$, $\gamma_{\mathrm{UL}}(t) = \lambda_{\mathrm{UL}}(t)/\mu_{\mathrm{UL}}$ and $\gamma_{\mathrm{DL}}(t) = \lambda_{\mathrm{DL}}(t)/\mu_{\mathrm{DL}}$. The call traffic load densities can be inferred from the historical load patterns [12, 34]. Since the large-scale variations over the day are finite, the slow timescale system presents a finite set of states.

Given the system state, the slow timescale action $W(t)$ specifies the BS mode of operation and the antenna tilting for the current period t. The action should not violate a target call-blocking probability for the uplink and downlink traffic, that is, it should satisfy $P_{\mathrm{BL}}^{\mathrm{UL}}(t) \leq \epsilon_{\mathrm{UL}}$ and $P_{\mathrm{BL}}^{\mathrm{DL}}(t) \leq \epsilon_{\mathrm{DL}}$, where $P_{\mathrm{BL}}^{\mathrm{UL}}(t)$ and $P_{\mathrm{BL}}^{\mathrm{DL}}(t)$ denote the call-blocking probabilities during period t in the uplink and downlink, respectively, and ϵ_{UL} and ϵ_{DL} are the target upper bounds in the uplink and downlink, respectively. Hence, the chosen action can be described by

$$W(t) = \{a_s(t) \forall s \in \mathcal{S} | P_{\mathrm{BL}}^{\mathrm{UL}}(t) \leq \epsilon_{\mathrm{UL}}, P_{\mathrm{BL}}^{\mathrm{DL}}(t) \leq \epsilon_{\mathrm{DL}}\}. \tag{7.17}$$

Two constraints are implicit in (7.17). First, all BSs in $\mathcal{S}$ cannot be switched off simultaneously. Second, active BSs must provide radio coverage for mobile users in inactive cells.

Given the system state $\Upsilon(t)$, the next state transition probability is independent of the action $W(t)$, that is,

$$\Pr(\Upsilon(t+1)|\Upsilon(t), W(t)) = \Pr(\Upsilon(t+1)|\Upsilon(t)). \tag{7.18}$$

In addition, the transition probability $\Pr(\Upsilon(t+1)|\Upsilon(t))$ is deterministic and inferred from the historical traffic patterns.

The slow timescale cost function is determined based on the fast timescale actions, a topic that will be addressed in the next subsection.

7.4.1.2   Fast Timescale

The fast timescale decision problem is to control the transmission powers of BSs and MTs to minimize (and balance) the total expected energy consumption. Since $\Lambda \ll \tau$, the fast timescale is an infinite horizon decision problem.

The fast timescale state $M(i, t)$ represents the number of mobile users with uplink and downlink traffic in the geographical region, that is, $M(i, t) = (m^{\mathrm{UL}}(i, t), m^{\mathrm{DL}}(i, t))$. The fast timescale system behaves as a discrete queuing system. Within period i, the arrivals of mobile users with uplink and downlink traffics follow Bernoulli processes with probabilities $\phi_{\mathrm{UL}} = \lambda_{\mathrm{UL}}(t)\Lambda/\tau$ and $\phi_{\mathrm{DL}} = \lambda_{\mathrm{DL}}(t)\Lambda/\tau$ in the uplink and downlink, respectively. The uplink and downlink service processes follow geometric distributions with parameters $\nu_{\mathrm{UL}} = \mu_{\mathrm{UL}}\Lambda/\tau$

and $\nu_{\mathrm{DL}} = \mu_{\mathrm{DL}}\Lambda/\tau$, respectively, where μ_{UL} and μ_{DL} are expressed in per unit τ. A maximum of $M_{\mathrm{UL}}(M_{\mathrm{DL}})$ MTs can be served simultaneously in the uplink (downlink), $M_{\mathrm{UL}} = \sum_s M_s^{\mathrm{UL}} 1_{a_s \neq 0}$ ($M_{\mathrm{DL}} = \sum_s M_s^{\mathrm{DL}} 1_{a_s \neq 0}$), where $1_{a_s \neq 0}$ denotes the indicator function. Hence, the uplink and downlink queues can be described by $Geo/Geo/M_{\mathrm{UL}}/M_{\mathrm{UL}}$ and $Geo/Geo/M_{\mathrm{DL}}/M_{\mathrm{DL}}$ queues, respectively [203]. Define $\varsigma_{\mathrm{UL}}(\hat{m}_{\mathrm{UL}}|m_{\mathrm{UL}})$ as the probability that $\hat{m}_{\mathrm{UL}}$ MTs have completed their uplink service out of m_{UL} MTs in the uplink. It follows that

$$\varsigma_{\mathrm{UL}}(\hat{m}_{\mathrm{UL}}|m_{\mathrm{UL}}) = \frac{m_{\mathrm{UL}}!}{\hat{m}_{\mathrm{UL}}! \, (m_{\mathrm{UL}} - \hat{m}_{\mathrm{UL}})!} \nu_{\mathrm{UL}}^{\hat{m}_{\mathrm{UL}}} (1 - \nu_{\mathrm{UL}})^{m_{\mathrm{UL}} - \hat{m}_{\mathrm{UL}}}. \tag{7.19}$$

Let $\Gamma(m^{\mathrm{UL}}(i,t))$ and $\Gamma(m^{\mathrm{DL}}(i,t))$ denote the steady-state probabilities of having $m^{\mathrm{UL}}(i,t)$ and $m^{\mathrm{DL}}(i,t)$ MTs in the uplink and downlink, respectively. The balance equations for the uplink queue are given by

$$\phi_{\mathrm{UL}}\Gamma(0) = \sum_{m=1}^{M_{\mathrm{UL}}} \Gamma(m)(1 - \phi_{\mathrm{UL}})\varsigma_{\mathrm{UL}}(m|m)$$

$$\Gamma(m) = \sum_{m'=m}^{M_{\mathrm{UL}}} \Gamma(m')[(1 - \phi_{\mathrm{UL}})\varsigma_{\mathrm{UL}}(m' - m|m') \tag{7.20}$$

$$+ \phi_{\mathrm{UL}}\varsigma_{\mathrm{UL}}(m' + 1 - m|m')] + \Gamma(m - 1)\phi_{\mathrm{UL}}\varsigma_{\mathrm{UL}}(0|m - 1),$$

$$1 \leq m \leq M_{\mathrm{UL}} - 1$$

$$\Gamma(M_{\mathrm{UL}}) = \Gamma(M_{\mathrm{UL}} - 1)\phi_{\mathrm{UL}}\varsigma_{\mathrm{UL}}(0|M_{\mathrm{UL}} - 1)$$

$$+ \Gamma(M_{\mathrm{UL}})[\phi_{\mathrm{UL}}\varsigma_{\mathrm{UL}}(1|M_{\mathrm{UL}}) + \varsigma_{\mathrm{UL}}(0|M_{\mathrm{UL}})].$$

Similar equations can be written for the downlink queue. Using the balance equations, one can easily derive the steady-state probabilities $\Gamma(m^{\mathrm{UL}})$ and $\Gamma(m^{\mathrm{DL}})$ and the transition probabilities $\Pr(\hat{m}^{\mathrm{UL}}|m^{\mathrm{UL}})$ and $\Pr(\hat{m}^{\mathrm{DL}}|m^{\mathrm{DL}})$ by solving a set of linear equations. In finding $W(t)$, the blocking probabilities $P_{\mathrm{BL}}^{\mathrm{UL}}(t)$ and $P_{\mathrm{BL}}^{\mathrm{DL}}(t)$ are given by $\Gamma(M_{\mathrm{UL}})$ and $\Gamma(M_{\mathrm{DL}})$, respectively, for the current time period arrival rates ϕ_{UL} and ϕ_{DL}.

Given the slow timescale action $W(t)$ and the fast timescale state $M(i,t)$, the fast timescale action $U(i,t)$ controls the transmission powers of the BSs and MTs, that is,

$$U(i,t) = (P_{\mathrm{DL},s}(a_s(t), m_s^{\mathrm{DL}}(i,t)), P_{\mathrm{UL},s}(a_s(t), m_s^{\mathrm{UL}}(i,t))|$$

$$P_{\mathrm{DL},s}(a_s(t), m_s^{\mathrm{DL}}(i,t)) \leq P_{s,\mathrm{mx}}, \tag{7.21}$$

$$P_{\mathrm{UL},s}(a_s(t), m_s^{\mathrm{UL}}(i,t)) \leq P_{\mathrm{UL},\mathrm{mx}}).$$

Given the system state $M(i,t)$, the next-state transition probability is independent of the action $U(i,t)$, and is given by

$$\Pr(M(i + 1,t)|M(i,t)) = \Pr(m^{\mathrm{UL}}(i + 1,t)|m^{\mathrm{UL}}(i,t))$$

$$\times \Pr(m^{\mathrm{DL}}(i + 1,t)|m^{\mathrm{DL}}(i,t)). \tag{7.22}$$

Given the actions taken at period i, the BSs' expected energy consumption is

$$E_{\mathrm{DL}}(m^{\mathrm{DL}}(i,t)) = \Gamma(m^{\mathrm{DL}}(i,t)) \sum_s P_{\mathrm{DL},s}(a_s(t), m^{\mathrm{DL}}(i,t)), \tag{7.23}$$

and the expected energy consumption for the MTs at the uplink is given by

$$E_{\mathrm{UL}}(m^{\mathrm{UL}}(i,t)) = \Gamma(m^{\mathrm{UL}}(i,t)) \sum_s m^{\mathrm{UL}}(i,t) P_{\mathrm{UL},s}(a_s(t), m^{\mathrm{UL}}(i,t)). \qquad (7.24)$$

Since we aim to minimize (and balance) the total expected energy consumption, the cost function is expressed as

$$O_f(M(i,t), U(i,t)) = E_{\mathrm{DL}}(m^{\mathrm{DL}}(i,t)) + \upsilon E_{\mathrm{UL}}(m^{\mathrm{UL}}(i,t)), \qquad (7.25)$$

where υ is a weighting factor.

We derive the optimal BS-switching decisions and transmission power control for MTs and BSs in the next subsection.

7.4.1.3 Optimal Decisions

Given the system state $\Upsilon(t)$ and action $W(t)$ in period t, the fast timescale decision framework total value function is given by

$$V_{\pi_t}(M_0(t)) = \mathbb{E}\left\{ \lim_{I\to\infty} \frac{1}{I} \sum_{i=1}^{I} O_f(M(i,t), U(i,t)) | W(t) \right\}, \qquad (7.26)$$

where $M_0(t)$ is the initial system state in period t and $\mathbb{E}$ denotes the expectation, which is taken over the system state $M(i,t)$ and π_t is the policy of the fast timescale decision problem in period t. Hence, π_t is a set of actions taken for all $i \in \mathcal{I}$ at a given t, with a policy space Π_t. The immediate cost of the slow timescale decision problem given the fast timescale policy in period t is expressed as

$$O_s(\Upsilon(t), W(t), \pi_t) = \mathbb{E}\{V_{\pi_t}(M_0(t))\}, \qquad (7.27)$$

where the expectation is taken over the initial system state $M_0(t)$. Let $\pi = \{W(1), \ldots, W(T)\}$ denote the slow time scale policy, which admits the policy space of Π. The dynamic planning approach with balanced energy efficiency follows the policies π and π_t for all $t \in \mathcal{T}$ and minimizes the total uplink and downlink expected energy cost, that is,

$$\min_{\pi \in \Pi} \min_{\pi_1, \pi_2, \ldots \pi_t \in \Pi_t} \mathbb{E}\left\{ \sum_{t=1}^{T} O_s(\Upsilon(t), W(t), \pi_t) \right\}, \qquad (7.28)$$

and the expectation is over the states $\Upsilon(t)$.

The optimal solution for (7.28) is a sequence $(W(t), \pi_t)$ for all $t \in \mathcal{T}$ that minimizes the expected total energy consumption expressed by Ismail et al. [198]

$$\pi_t = \arg\min_{\pi_t \in \Pi_t} \mathbb{E}\left\{ \lim_{I\to\infty} \frac{1}{I} \sum_{i=1}^{I} O_f(M(i,t), U(i,t)) | W(t) \right\}$$

$$W(t) = \arg\min_{W(t)} O_s(\Upsilon(t), W(t), \pi_t). \qquad (7.29)$$

The approach is to find the optimal fast timescale policy π_t that minimizes the expected total energy consumption, given some action $W(t)$ taken at the slow timescale. For different actions $W(t)$, different MT associations can be deployed, leading to different cost values $O_f(M(i,t), U(i,t))$. Therefore, in the second step, we determine the optimal slow timescale action $W(t)$ that minimizes the expected total energy consumption $O_s(\Upsilon(t), W(t), \pi_t)$.

7.4.2   Performance Evaluation

In this subsection, we evaluate the performance of the proposed dynamic planning approach with balanced energy efficiency through comparison with a traditional dynamic planning approach that does not account for the energy consumption of the mobile users (i.e. with $v = 0$). The system model is given in Figure 7.11. The two BSs are identical and the system parameters are given by $Y_s = 100$ m, $d_{1,1} = 150$ m, $\phi_{\mathrm{UL}} = \phi_{\mathrm{DL}} = 0.01$, $M_{\mathrm{UL}} = M_{\mathrm{DL}} = 7$, $R_{\mathrm{UL}} = R_{\mathrm{DL}} = 5$ Mbps, $B_s^{\mathrm{UL}} = B_s^{\mathrm{DL}} = 5$ MHz, $P_{F,s} = 390$ W, $\Delta_s = 4.7$ and $P_{0,\mathrm{s}} = 75$ W, $\tau = 1$ h, $\Lambda = 5$ min and $v = 50$. The fast time scale arrival rate in the downlink is 0.5, and the average service duration is 0.2 for both the uplink and downlink.

Figure 7.13a and b show the expected downlink energy consumption for both balanced and unbalanced dynamic planning. The unbalanced dynamic planning energy consumption performance does not vary with the weighting factor v since it does not account for the

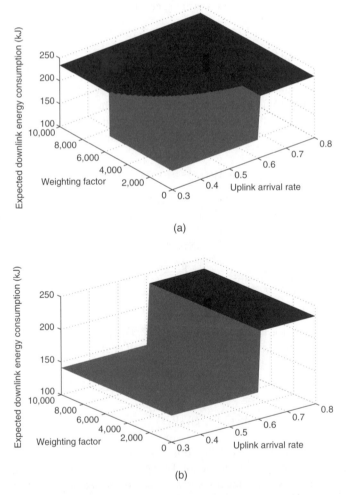

(a)

(b)

Figure 7.13   Expected downlink energy consumption versus the arrival rate of uplink users and the weighting factor v [197]: (a) balanced approach and (b) unbalanced approach. The spatial distribution is $\rho_{\mathrm{DL}}(A_{1,1}) = 0.8$ for downlink users

MTs' incurred energy consumption. It is affected only by the arrival rate. At low arrival rates $[0.3, 0.6]$, only one BS is kept active to serve the MTs, while at higher arrival rates higher than 0.6, both BSs are switched on to satisfy the target service quality in terms of minimum required data rates (and hence upper bound on call-blocking probabilities). On the contrary, the balanced dynamic planning energy consumption performance is affected by both v and arrival rate. For low arrival rates and low v, a single BS is kept active to serve the MTs. As the arrival rate increases, a second BS is switched on to satisfy the users target service quality (in terms of minimum required data rate and call-blocking probabilities). In addition, large v values force the second BS activation to avoid uplink service degradation (e.g. a higher call-dropping rate and lower throughput) due to MTs' battery depletion. At low arrival rate values, the second BS activation is dominated by large v values since the uplink service degradation due to MTs' battery depletion is more pronounced than users' call blocking due to limited radio resources, and the opposite is true at high arrival rate values. In Figure 7.14a and b, when a single BS

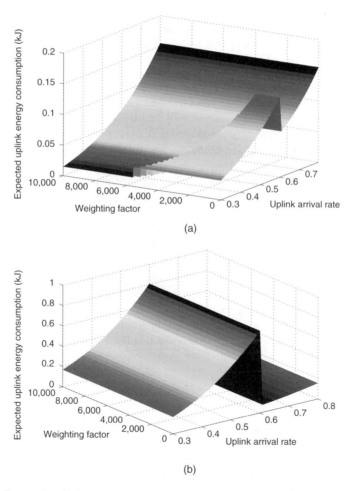

(a)

(b)

Figure 7.14 Expected uplink energy consumption versus the arrival rate of uplink users and the weighting factor v [197]: (a) balanced approach and (b) unbalanced approach. The spatial distribution is $\rho_{\text{UL}}(A_{2,1}) = 0.7$ for uplink users

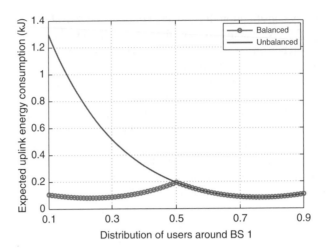

Figure 7.15 Expected energy consumption of uplink users versus the spatial distribution of the uplink users near the proximity of the first BS [197]. The uplink users' arrival rate is 0.4

is switched on at arrival rates $[0.3, 0.6]$ and $v = 5{,}550$ to $1{,}400$, respectively, the balanced approach decides which BS should be kept active based on the spatial distribution of the uplink users. Hence, for the balanced approach, the second BS is kept active while the first BS is switched off. However, for the unbalanced approach, the expected energy consumption of the uplink users is not accounted for, and hence, the first BS is kept active while the second BS is switched off. Even if the unbalanced approach follows a random or round-robin BS switching off policy to decide which BS should be switched off, the unbalanced approach still will lead to a higher expected uplink energy consumption compared with the balanced approach.

Figure 7.15 shows the expected uplink energy consumption versus the spatial distribution of uplink users near the proximity of the first BS. The arrival rate for the uplink users is fixed at 0.4. Due to the low arrival rate, only a single BS is kept active (Figure 7.13a). As shown in Figure 7.15, with more uplink users concentrated around the second BS ($\rho_{UL}(A_{1,1}) \in [0.1, 0.5]$), the balanced dynamic planning approach keeps the second BS active and switches off the first BS, resulting in low expected energy consumption for the uplink users, unlike the unbalanced approach which keeps the first BS active and switches off the second BS. As uplink users become more concentrated around the first BS ($\rho_{UL}(A_{1,1}) > 0.5$), the balanced approach switches off the second BS and keeps the first BS active to keep the expected energy consumption of the uplink users as minimum as possible.

7.5 Summary

At a low call traffic load, network operators are expected to save energy by switching off some of their BSs. In this context, two approaches can be adopted to serve the mobile users. The first approach relies on the same network resources, and hence, a dense macro–pico architecture is implemented and optimal cell zooming techniques (i.e. expanding and shrinking the cell size) and sleep policies (i.e. on–off switching of BSs) for both macro and pico BSs are deployed depending on the traffic pattern. In such a case, it is imperative to satisfy the users' target QoS while saving energy for the network operator. The second approach that can be adopted to serve the mobile users while switching off lightly loaded BSs relies on cooperative networking, where network operators with overlapped coverage among different BSs alternately switch on and off their BSs to achieve energy saving and avoid the dynamic planning shortcomings as mobile users are served by the active network operator. In all cases, it is beneficial to combine both switching off BSs and switching off some of the radio resources of an active BS to further improve the amount of energy saving. Finally, dynamic planning should capture and balance the trade-off in energy efficiency between network operators (in the downlink) and mobile users (in the uplink) to save energy for network operators while not jeopardizing the mobile users' perceived service quality due to a high rate battery depletion for MTs which results in a high rate of dropped services in the uplink.

8

Greening the Cell Edges

Heterogeneous networks (HetNets) are envisioned to enable the next-generation cellular networks with higher spectral and energy efficiency. This chapter presents a two-tier HetNet, where small-cell BSs (SBSs) are arranged around the edge of the reference macro-cell such that the resultant configuration is referred to as cell-on-edge (COE). Each mobile user in a small-cell is considered to be capable of adapting its uplink transmission power according to a location-based slow power control (PC) mechanism. The COE configuration is observed to increase the uplink area spectral efficiency (ASE) and energy efficiency, while reducing the co-channel interference power. A moment-generating function (MGF)-based approach is presented to derive the analytical bounds on the uplink ASE of the COE configuration. The derived expressions are generalized to any composite fading distribution and closed-form expressions are presented for the generalized-$\mathcal{K}$ fading channels. Simulation results are included to support the analysis and show the spectral and energy efficiency improvements of the COE configuration. A comparative performance analysis is also provided to demonstrate the improvement in the performance of cell-edge users of the COE configuration compared to macro-only networks and other unplanned deployment strategies. Moreover, the COE deployment guarantees the reduction in carbon footprint of the mobile operations by employing adaptive uplink power control. In order to calibrate the reduction in CO_2 emissions, this chapter provides a description of the ecological and associated economical impacts of energy savings in the proposed deployment.

8.1 Introduction

The telecommunications industry is currently witnessing a remarkable increase in the data and voice traffic, particularly with the introduction of smartphones, tablet computers and other portable smart devices. Today, the mobile data volume corresponding to 4.5 billion mobile subscribers (which represents 67% of the world population) is 45 million TB/year, and is expected to reach 623 million TB/year by 2020. The exponential growth in the data rates is due to not only an increase in the number of broadband mobile subscribers, but also bandwidth-consuming activities such as distribution of videos, online meetings, e-government services and facilities and other peer-to-peer information and content exchange services. A direct solution is to densify the BS deployment. Adding a BS to the sparsely deployed areas does not have much impact on the interference, and thus cell splitting gains are easy to achieve. Nonetheless, adding a BS to the densely deployed urban area generates severe interference per channel, and therefore, the cell-splitting gains get reduced significantly. In addition, the site acquisition cost in a capacity-limited dense urban area may also get prohibitively expensive.[1]

The demand for high data rates is also expected to increase (i) the network energy consumption by 16–20%; (ii) network operational expenses by 20–30% and (iii) carbon footprint of mobile communication industry by 10%, by 2020 [9, 60, 205–207]. The contributing factors behind the increase in energy consumption and carbon footprint include, but are not limited to, production, operation, distribution and maintenance of the mobile communications networks, devices and services. Therefore, the wireless network operators are facing a huge challenge to meet the escalated demands of mobile users while minimizing the energy consumption and cost of the wireless networks.

In this regard, heterogeneous networks are emerging as the most influential solution that guarantees higher data rates, offloading of macro-cell traffic and reduction of CAPEX and OPEX, while providing dedicated capacity to homes, enterprises or urban hot spots. In addition to macro-cell networks, HetNets include various kinds of small-cells, such as outdoor/indoor femto cells, relays and micro and pico cells, with radii of about 30–200 m. The small-cells are short-range, low-power and low-cost BSs that guarantee spectral and energy efficiency by reducing the propagation distance between the BS and the mobile users in the small-cells [208–212].

8.1.1 Why Cell-on-Edge Deployment?

In LTE-Advanced, the focus is on increased peak data rates, higher spectral and energy efficiency, increased number of simultaneously active mobile users and improved performance at the cell-edges. The spectral efficiency of the cell-edge mobile users is often very poor due to the higher path-loss effects and thereby degrades the overall network coverage and capacity. Due to this reason, the network operators are striving to facilitate the cell-edge mobile users in a cost-effective manner. Several small-cell deployment strategies are currently under consideration where the performance is calibrated with respect to the profitability, spectral and energy efficiency, link quality and outage probability [213–216]. However, providing coverage to the cell-edge mobile users in a spectral and energy-efficient manner is still a challenge.

Limitations in cellular coverage, particularly at the cell-edge, can be overcome by positioning small-cell BSs at the cell-edge or in a coverage hole to compensate for the drastic path-loss experienced at the cell-edge. In this context, this chapter focuses on the cell-edge deployment of the small-cells by the operator and the resultant configuration is referred to as COE configuration. The main aim of this deployment strategy is to provide the required network coverage and capacity for the cell-edge mobile users by reducing the distance between the transmitting BS and the cell-edge mobile users in an energy-efficient manner.

---

[1] As an example, the total capital expenditure (CAPEX) of a macro BS equals Euro 0.02 million. Whereas, the annual average operational expenditure (OPEX) per site is approximately equal to the electricity bill, which is given as Euro 0.1 per kilowatt-hour. If a typical macro BS annually consumes 8,000 kWh, then the annual OPEX of the site can be evaluated as Euro 800 [21, 204].

The COE configuration is expected to produce significant gains compared with two competitive network configurations, namely (i) HetNets, where the small-cells are uniformly distributed across the macro-cells, that is, uniformly distributed small-cells (UDC) and (ii) macro-only network (MoNet). Typically, UDC is considered as one of the standard approaches that allow unplanned deployment of the small-cells in the current infrastructure [217–219]. Even though, considering the UDC deployment may be more close to realistic deployments, the considered COE deployment is simple, easy to assess, analytically tractable, helps to conduct rapid performance assessment studies and excels UDC in the following aspects:

- **Energy Consumption:** Due to the limited battery power constraint and uplink power control, mobile users located close to the serving BS can achieve their desired signal to interference ratio (SINR) while minimizing their transmission power. Whereas, the cell-edge mobile users are highly likely to transmit with their maximum powers to achieve as high data rates as possible. In this context, the COE deployment reduces the transmission power of the cell-edge mobile users while allowing them to achieve their desired signal quality levels.
- **Spectral Efficiency:** The spectral efficiency of the HetNets increases with the increase of deployment density of small-cells. The small-cell deployment guarantees the reduction in path-loss between the mobile user and the small-cell BS, and thereby improves the link quality. The uplink spectral efficiency improvement is expected to be significant in the COE deployment, as the absence of such a configuration forces the cell-edge mobile users to get connected to the macro-cell BS via poorer channel conditions.
- **Interference Reduction:** In the case of a MoNet, the cell-edge mobile users are highly likely to transmit with their maximum power to achieve and maintain the desired SINR. The co-channel interference due to such cell-edge mobile users may cause significant degradation in the network performance with aggressive frequency reuse. The COE configuration enables a reduction of transmission power of the cell-edge mobile users due to the reduction in terms of distance between the transmitter and receiver, which in turn diminishes the co-channel interference with limited spectral resources.

8.1.2 Background Work

Some recent simulation-based studies investigated the performance of micro cell and pico cell deployments in terms of downlink area power consumption and spectral efficiency metrics [218–220]. The investigations in [218, 219] have shown that the power savings from the deployment of micro BSs at the macro cell-edges are moderate in full-load scenarios and strongly depend on the offset power consumption of both macro and micro sites. It has been further shown in [220] that the deployment of residential pico cells can reduce the total network energy consumption by 60% in urban areas. Nonetheless, it is important to note that all of these studies are focused on downlink performance analysis, and there is no explicit framework that investigates uplink spectral and energy efficiency of HetNets with small-cells on the edge. Since the uplink power consumption is directly related to the user's channel conditions, the type of uplink power control employed and battery power constraint, and therefore, the conclusions for downlink networks may not be directly applicable to the uplink scenarios. Moreover, it is also worth mentioning that the deployment of small-cells does not directly lead to a reduction in power consumption, rather it depends on the type of uplink power control employed and required quality of service (QoS) at mobile user's end. The higher QoS and traffic loads certainly lead to higher power consumption and consequently results in reduced power efficiency.

8.2   Two-Tier Small-Cell-on-Edge Deployment

This section presents the network layout, bandwidth partition and random mobile user distribution.

8.2.1   Network Layout

Consider a two-tier HetNet as illustrated in Figure 8.1, where the integration of macro-cell and small-cell networks is illustrated. For the sake of simplicity, the considered configuration has only the central macro-cell of the macro-cell network tier. However, it is assumed that there are $M - 1 = 6$ interfering co-channel macro-cells near the reference macro-cell. The first tier of the considered HetNet comprises circular macro-cells, each of radius R_m [m] with a BS B_m deployed at the centre and equipped with an omni-directional antenna. Each macro-cell is assumed to have U_m mobile users uniformly distributed over the region bounded by R_0 and R_m, where R_0 denotes the minimum distance between the macro-cell mobile user and its serving BS.

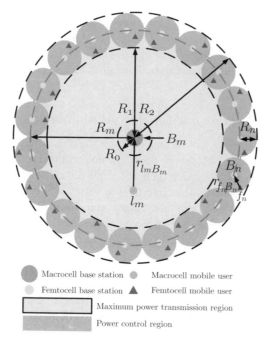

Figure 8.1   Graphical illustration of the two-tier HetNets, where a macro-cell is surrounded by N small-cells around the edge of the reference macro-cell

The second tier of the HetNet consists of N circular small-cells (e.g. outdoor femto cells) each of radius R_n [m] with low-power low-cost operator deployed BSs B_n located at the centre of each small-cell. It is considered that the small-cells are distributed around the edge of the reference macro-cell. The resultant small-cell deployment is referred to as COE configuration. For practical reasons, the number of small-cells per macro-cell can be calculated as follows:

$$
N = \begin{cases} \left\lceil \mu \dfrac{(R_2^2 - R_1^2)}{R_n^2} \right\rceil = \left\lceil \mu \dfrac{4R_m}{R_n} \right\rceil & R_m > R_n \\ 0 & R_m \leq R_n, \end{cases}
\tag{8.1}
$$

where $R_1 = R_m - R_n$, $R_2 = R_m + R_n$, $\lceil \cdot \rceil$ denotes the ceiling function and the factor μ $(0 < \mu \leq 1)$ is referred to as the cell population factor (CPF), which controls the number of small-cells per macro-cell, that is,

$$
\mu = \begin{cases} 0 & \text{off-load small-cells}^2 \\ 1 & \text{maximum number of small-cells per macro-cell.} \end{cases}
\tag{8.2}
$$

The number of mobile users in each small-cells is expressed as $U_n = (U - U_m)/N$, where $U_m = U(R_1^2 - R_0^2)/R_m^2$ and $U = U_m + NU_n$. In the COE deployment, U_m out of U mobile users are assumed uniformly distributed over the region bounded by R_0 and R_1, whereas the remaining mobile users, that is, $U - U_m$, are reserved for N small-cells. Moreover, the frequency allocated to a reference macro-cell is reused in the neighbouring macro-cells at a reuse distance $D' = R_u(R_m + R_n)$ [m], where R_u represents the network resource reuse factor. Here, the reuse distance is calculated based on the circular coverage of the macro-cells such that the network resource reuse factor is given by $R_u = 2\sqrt{k}$, where k is the conventional frequency reuse factor given by $k = \hat{i}^2 + \hat{i}\hat{j} + \hat{j}^2$. An aggressive frequency reuse factor, $k = 1$ has been assumed for the pair $(\hat{i}, \hat{j}) = (1, 0)$ [222]. The total bandwidth allocated to the small-cell tier is reused in each small-cell within a given macro-cell.

8.2.2   *Bandwidth Partition and Channel Allocation*

The spectrum partition strategy will be employed for the considered HetNets, which include COE and UDC configurations. Moreover, the spectrum partition is based on the proportion of the number of mobile users in the macro-cell and small-cells. The spectrum-splitting strategy has been considered to avoid cross-tier interference issues, that is, the interference between macro-cells and small-cells. However, this is not a limitation as it can be applied to spectrum sharing scenarios as well, by conducting a more comprehensive mathematical analysis. If w_t [Hz] is the total bandwidth of the available spectrum per cell, then the total bandwidth may be divided as

$$
w_t = w_m + w_n,
\tag{8.3}
$$

where $w_m = w_t(U_m/U)$ [Hz] and $w_n = w_t(NU_n/U)$ [Hz] represent the amount of the spectrum dedicated to the macro-cell and small-cells, respectively, based on the proportion of active

---

[2] Here, off-load small-cells refer to the mechanism, where small-cells are inactive. As an example, in [221], it is shown that the downlink power consumption of HetNets can be reduced significantly by forcing the small-cells to turn on the sleeping mode during low- and medium-traffic load conditions by controlling the CPF.

mobile users. The macro-cell and small-cell bandwidth are divided further into sub-channels, where each sub-channel can be allocated to one mobile user at a time and there will not be any mobile user, who cannot be serviced by the respective macro-cell or small-cell BS. The number of active serviced channels available per macro-cell and small-cells can then be expressed as $N_m = w_m/U_m$ and $N_n = w_n/U_n$, respectively.[3] Each sub-channel is allocated to any user randomly without considering the channel conditions, that is, strictly a fair scheduling strategy is considered.

8.2.3   Mobile User Distribution

All of the mobile users in macro-cell and small-cell networks are considered as mutually independent and uniformly distributed in their respective cells. The PDF of the location of a macro-cell mobile user located at (r, θ) from the serving macro-cell BS is expressed as

$$f_r(r) = \frac{(r - R_0)}{R_1^2 - R_0^2}, \ f_\theta(\theta) = \frac{1}{2\pi}, \tag{8.4}$$

where $R_0 \leq r \leq R_1$ and $0 \leq \theta \leq 2\pi$. Similarly, the PDF of the location of a small-cell mobile user located at $(\tilde{r}, \tilde{\theta})$ from the serving SBS can be expressed as

$$f_{\tilde{r}}(\tilde{r}) = \frac{2\tilde{r}}{R_n^2}, \ f_{\tilde{\theta}}(\tilde{\theta}) = \frac{1}{2\pi}, \tag{8.5}$$

where $0 \leq \tilde{r} \leq R_n$ and $0 \leq \tilde{\theta} \leq 2\pi$ (see Fig. 8.1 for a geometrical representation of R_2 and R_1).

8.3   Energy-Aware Transmission Design

The radio environment of a typical wireless cellular network is subject to (i) distance-dependent path-loss, (ii) shadowing and (iii) multipath fading. The radio wave propagation in small-cells is complicated due to strong LOS conditions between the transmitter and receiver. Several models can be employed for this purpose [223]. However, it has been shown that a simple path-loss model does not fit well the measurements for strong LOS environments [224]. Motivated by this fact, this chapter considers a two-slope (or commonly known as dual-slope) path-loss model, which is shown to be suitable for strong LOS conditions [223].

8.3.1   Path-Loss Model for Strong LOS Conditions

The dual-slope path-loss model considers two separate path-loss exponents α_1 and α_2, which are referred to as basic and additional path-loss exponents, respectively. These path-loss exponents are used to characterize two different propagation environments, together with a break-point distance g between them, where propagation changes form one regime to the other. The

---

[3] In practical cases, the number of active serviced channels available per macro-cell and small-cell can be expressed as $N_m = w_m/\Delta$ and $N_n = w_n/\Delta$, respectively, where Δ can be selected arbitrarily.

signal attenuates with the basic path-loss exponent α_1 before break point and attenuates with the additional path-loss exponent α_2 after break point. For $r \leq g$, the path-loss can be modelled as $1/L = K/r^{\alpha_1}$, whereas for $r > g$, it can be modelled as $1/L = K/(r/g)^{\alpha_2}g^{\alpha_1}$, where $g = 4h_{\text{rx}}h_{\text{tx}}/\lambda_c$ [m] is the break point of a path-loss curve and it depends on the macro-cell or small-cell BS's (receiver in uplink) antenna height h_{rx} [m], the antenna height of the mobile user (transmitter in uplink) h_{tx} [m] and wavelength of the carrier frequency λ_c. Constant K represents the path-loss constant.

The dual-slope path-loss model can be written in a generalized form as $1/L = K/r^{\alpha_1}(1 + r/g)^\beta$, where $\beta = \alpha_2 - \alpha_1$.

The received signal power at macro-cell or small-cell BS from the corresponding mobile user is expressed as

$$P^{\text{rx}} = \frac{K}{r^{\alpha_1}(1 + r/g)^\beta}P^{\text{tx}}\zeta, \tag{8.6}$$

where P^{rx} [W] denotes the average received signal power at the reference macro-cell or small-cell BS from the desired mobile user, which is located at a distance of r from the same reference BS, ζ is the composite shadowing and fading over the link between the mobile user and respective macro-cell or small-cell BS and P^{tx} [W] defines the mobile user transmission power, which is equal to the maximum power P_{max} for a macro-cell user and is expressed for a small-cell user according to the slow power control (PC) mechanism as follows [225–228]:

$$P^{\text{tx}} = \min\left(P_{\text{max}}, P_0 \frac{r^{\alpha_1}(1 + r/g)^\beta}{K}\right), \tag{8.7}$$

where P_0 is the cell-specific parameter and it is used to control the target SINR. Using (8.7) and (8.6) can be expressed as

$$\begin{cases} P^{\text{rx}} = P_{\text{max}}\frac{K}{r^{\alpha_1}(1+r/g)^\beta}\zeta & P_{\text{max}} < P_0\frac{r^{\alpha_1}(1+r/g)^\beta}{K} \\ P_0\zeta & \text{otherwise.} \end{cases} \tag{8.8}$$

8.3.2 Composite Fading Channel for Strong LOS Conditions

In wireless channels, the phenomena of shadowing and fading can be jointly modelled by the composite fading distribution. Nakagami-m is a generic fading distribution, which includes Rayleigh distribution for $m = 1$ (typically used for non-LOS conditions) and can well approximate the Ricean fading distribution for $1 \leq m \leq \infty$ (typically used for strong LOS conditions) [229, 230]. Shadowing is usually modelled by a log-normal distribution. However, due to the unavailability of a closed-form expression, log-normal-based composite fading models further complicate the analysis. Recently, it has been shown that the Nakagami log-normal distribution can be modelled by the generalized-$\mathcal{K}$ distribution, where the average power variations due to shadowing are closely approximated by gamma distribution [231].

Figure 8.2 shows the summary of uplink transmission power per mobile user over the range of desired target SINR for HetNets and other competitive networks, namely MoNets with and without PC, HetNets with UDC deployment and HetNets with COE configuration. The mobile users in traditional MoNets without PC transmit with the maximum power over the link, while the mobile users in MoNets with PC transmit with the minimum required power to meet

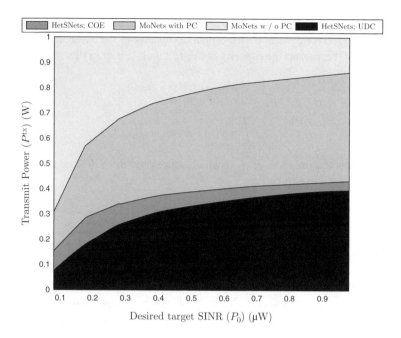

Figure 8.2 Summary of uplink transmission power adaptation for several competitive networks configurations

the desired SINR. Similarly, the mobile users in HetNets adapt their power intelligently and transmit with the minimum power required to meet the quality of the link. The adaptive mobile user transmission power in HetNets represents the average of the minimum transmission power of the macro-cell and small-cell mobile users. The transmission power of the mobile user increases with the rise in the desired target SINR. The reduction in transmission power due to PC is significant in HetNets due to the shorter distances. Moreover, the power consumption of the HetNets with COE deployment is lower than that corresponding to the UDC deployment, since under the UDC deployment mobile users are located around the edge of the cell while transmitting with their maximum power.

8.4 Area Spectral Efficiency of HetNets

ASE η of typical macro-cell and small-cell networks is mathematically defined as follows [222, 232]:

$$\eta = \frac{C_h}{\pi w_h (D'/2)^2} = \frac{4C_h}{\pi w_h R_u^2 (R_m + R_n)^2},\tag{8.9}$$

where $D' = R_u(R_m + R_n)$, R_u denotes the frequency reuse factor and C_h denotes the total achievable Shannon capacity of two-tier HetNets, which is given by

$$C_h = C_m + C_n = \sum_{l=1}^{U_m} C_{l_m} + \sum_{n=1}^{N} \sum_{f=1}^{U_n} C_{f_n},\tag{8.10}$$

where C_m and C_n [bps/Hz] are the mean achievable capacity of the mth macro-cell and N small-cells, respectively, C_{l_m} denotes the Shannon capacity of the lth mobile user in the mth macro-cell and C_{f_n} represents the capacity of the fth mobile user in the nth small-cell. More explicitly, C_{l_m} is given by

$$C_{l_m} = w_{l_m} \, \mathbb{E}[\log_2(1 + \gamma_{l_m})], \tag{8.11}$$

$$= w_{l_m} \int_0^\infty \log_2(1 + \gamma_{l_m}) f_\gamma(\gamma_{l_m}) \mathrm{d}\gamma_{l_m}, \tag{8.12}$$

where $f_\gamma(\gamma_{l_m})$ denotes the PDF of γ_{l_m}, which represents the SINR of the lth macro-cell mobile user in the mth macro-cell:

$$\gamma_{l_m} = \frac{P^{\mathrm{rx}}_{l_m, B_m}(r_{l_m, B_m})}{\sum\limits_{i=1, i \neq m}^{M} P^{\mathrm{rx}}_{l_i, B_m}(r_{l_i, B_m}) + \sigma^2}. \tag{8.13}$$

In (8.13), σ^2 denotes the thermal noise power, $P^{\mathrm{rx}}_{l_m, B_m}(r_{l_m, B_m})$ denotes the received power level at the reference macro-cell BS B_m from the lth desired mobile user and $\sum_{i=1, i \neq m}^{M} P^{\mathrm{rx}}_{l_i, B_m}(r_{l_i, B_m})$ represents the sum of the individual interfering power levels received at the reference macro-cell BS B_m from the interfering mobile users $\{l_i\}_{i=1, i \neq m}^{M}$, which are located in each of the interfering macro-cell. As an example, the geometrical illustration of the macro-cell-level interference model is shown in Figure 8.3. Substituting (8.6) into[4] (8.13), the SINR of the macro-cell mobile user can be re-written as

$$\gamma_{l_m} = \frac{P_{\max} \zeta_{l_m, B_m} r_{l_m, B_m}^{-\alpha_{1m}} (g_m + r_{l_m, B_m})^{-\beta_m}}{\sum_{i=1, i \neq m}^{M} P_{\max} \zeta_{l_i, B_m} r_{l_i, B_m}^{-\alpha_{1m}} (g_m + r_{l_i, B_m})^{-\beta_m} + \sigma^2}, \tag{8.14}$$

where ζ_{l_i, B_m} is the composite fading statistics of the interference from the lth mobile user in the ith macro-cell to the mth BS of interest.

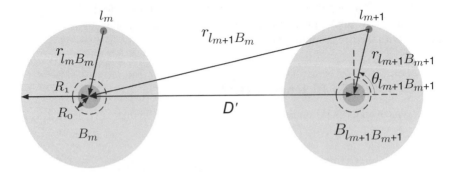

Figure 8.3 Geometrical illustration of the macro-cell-level interference problem, where the interfering mobile user is located at $(r_{l_{m+1}B_{m+1}}, \theta_{l_{m+1}B_{m+1}})$ in one of the M co-channel macro-cells at a reuse distance $D' = R_u(R_m + R_n)$

---

[4] In (8.6), the notations have been suppressed for the sake of simplicity and better understanding. For instance, $P^{\mathrm{rx}}_{l_m, B_m}(r)$, $P^{\mathrm{tx}}_{l_m, B_m}(r)$, ζ_{l_m, B_m}, α_{1m}, β_m and g_m are denoted by P^{rx}, P^{tx}, ζ, α_1, β and g, respectively.

Similarly, for the second tier of small-cells, C_{f_n} in (8.10) can be expressed as

$$C_{f_n} = w_{f_n} \mathbb{E}[\log_2(1 + \gamma_{f_n})] \tag{8.15}$$

$$= w_{f_n} \int_0^\infty \log_2(1 + \gamma_{f_n}) f_\gamma(\gamma_{f_n}) d\gamma_{f_n}, \tag{8.16}$$

where γ_{f_n} denotes the SINR of the fth mobile user located in the nth small-cell, which is given by

$$\gamma_{f_n} = \frac{P_{f_n,B_n}^{\mathrm{rx}}(\tilde{r}_{f_n,B_n})}{\sum_{j=n\pm1} P_{f_j,B_n}^{\mathrm{rx}}(\tilde{r}_{f_j,B_n}) + \sigma^2}, \tag{8.17}$$

where $P_{f_n,B_n}^{\mathrm{rx}}(\tilde{r}_{f_n,B_n})$ denotes the received power level at the reference small-cell BS B_n from the fth desired mobile user and $\sum_{j=n\pm1} P_{f_j,B_n}^{\mathrm{rx}}(\tilde{r}_{f_j,B_n})$ denotes the sum of the individual interfering power levels received at the reference small-cell B_n from the interfering mobile users located in the jth interfering macro-cell. Figure 8.4 illustrates the geometrical representation of the considered small-cell interference, where the interfering signals are considered from the mobile users $f_{n\pm1}$ located in two adjacent small-cells. However, this effect is considered only for analytical tractability and it does not affect the overall significance of COE configuration as it will be shown later through simulation results. Substituting (8.6) into[5] (8.17), γ_{f_n} can be expressed as

$$\gamma_{f_n} = \frac{P_{f_n,B_n}^{\mathrm{tx}} \zeta_{f_n,B_n} \tilde{r}_{f_n,B_n}^{-\alpha_{1n}} (g_n + \tilde{r}_{f_n,B_n})^{-\beta_n}}{\sum_{j=n\pm1} P_{f_j,B_j}^{\mathrm{tx}} \zeta_{f_j,B_n} \tilde{r}_{f_j,B_n}^{-\alpha_{1n}} (g_n + \tilde{r}_{f_j,B_n})^{-\beta_n} + \sigma^2}. \tag{8.18}$$

Figure 8.5 shows the ASE of four different types of network configurations: (i) MoNet (solid curve with triangle markers); (ii) COE configuration with interference from two adjacent small-cells (solid curve with square markers); (iii) COE configuration with interference from $N - 1$ small-cells (dotted curve with square markers) and (iv) UDC configuration with interference from $N - 1$ small-cells (solid curve with circle markers). The interference from $M - 1 = 6$ co-channel macro-cells is also considered in each of these configurations. It is clear that the ASE of the COE configuration has been significantly improved when the small-cells

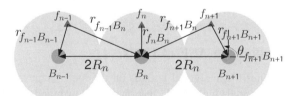

Figure 8.4 Geometrical illustration of the small-cell-level interference problem where the interfering mobile users are located at $(r_{f_{n\pm1}B_{n\pm1}}, \theta_{f_{n\pm1}B_{n\pm1}})$, that is, mobile users are located in two adjacent small-cells of the reference small-cell

---

[5] Here, the small-cell version of (8.6) can be obtained by replacing P^{rx}, P^{tx}, ζ, α_1, β and g by $P_{f_n,B_n}^{\mathrm{rx}}(\tilde{r})$, $P_{f_n,B_n}^{\mathrm{tx}}(\tilde{r})$, ζ_{f_n,B_n}, α_{1n}, β_n and g_n, respectively.

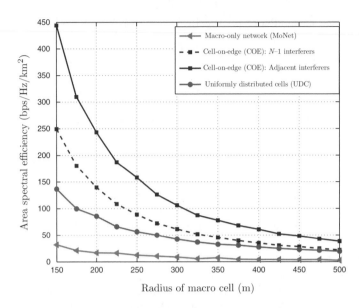

Figure 8.5 Comparison of the ASE of MoNet with two different HetNet configurations: (i) COE configuration and (ii) UDC configuration as a function of the reference macro-cell

are active in the macro-cell compared with the MoNet and UDC configurations (compare the solid curve with triangle markers with the rest of the curves with circle and square markers). This is due to the fact that the COE deployment restricts only the cell-edge mobile users to communicate with the small-cells, which enhances the overall network ASE compared with UDC and MoNet configurations. More precisely, the UDC configuration allows the small-cell BSs to be deployed in the cell centre, which causes an under-utilization of the macro-BS capabilities (under-utilization of existing infrastructure). Thus, the performance degradation due to the cell-edge mobile users still exist in the UDC configuration. Due to the weaker channel gains of the mobile users in the large macro-cells, the degradation of ASE with R_m is also evident.

8.5 Analytical Bounds on ASE of HetNets

This section first calculates the analytical bounds on the mean achievable capacity of COE deployment and then it proceeds with the derivation of the lower and upper bounds on ASE.

8.5.1 Mean Achievable Capacity Based on MGF Approach

The dependence of the distribution of SINR on the distribution of the interfering user locations and their fading channels leads to multifold convolutions. Recently, an MGF-based generalized framework has been developed in [233] to evaluate the system capacity, given the MGF of the desired signal and interference random variables. Using the efficient capacity lemma, the exact capacity of a desired macro-cell mobile user can be evaluated as follows:

$$C_{l_m} = w_{l_m} \int_0^\infty \underbrace{\int \cdots \int}_{M\,\text{Integrals}} \frac{\mathcal{M}_{I_m}(t) - \mathcal{M}_{S_m, I_m}(t)}{t} e^{-\sigma^2 t} f_{r_{l_m}, B_m}(r)$$

$$\cdot \prod_{i=1, i \neq m}^{M} f_{r_{l_i}, B_m}(r) \mathrm{d}r_{l_i, B_m} \, \mathrm{d}r_{l_m, B_m} \, \mathrm{d}t, \tag{8.19}$$

where $\mathcal{M}_{I_m}(t)$ and $\mathcal{M}_{S_m, I_m}(t) = \mathcal{M}_{S_m}(t)\mathcal{M}_{I_m}(t)$ denote the MGF of the macro-cell interference and joint MGF of the received signal and interference, respectively. In particular, $\mathcal{M}_{I_m}(t)$ and $\mathcal{M}_{S_m}(t)$ can be expressed as $\mathcal{M}_{I_m}(t) = \mathbb{E}[e^{-t \sum_{i=1, i \neq m}^{M} P_{l_i, B_m}^{\text{rx}}(r_{l_i, B_m})}]$ and $\mathcal{M}_{S_m}(t) = \mathbb{E}[e^{-t P_{l_m, B_m}^{\text{rx}}(r_{l_m, B_m})}]$, respectively. Similarly, for small-cell networks, the capacity of a desired small-cell user can be expressed as

$$C_{f_n} = w_{f_n} \int_0^\infty \underbrace{\int \cdots \int}_{\text{Three Integrals}} \frac{\mathcal{M}_{I_n}(t) - \mathcal{M}_{S_n, I_n}(t)}{t} e^{-\sigma^2 t} f_{\tilde{r}_{f_n}, B_n}(\tilde{r})$$

$$\times \prod_{j \neq 1} f_{\tilde{r}_{f_j}, B_n}(\tilde{r}) \mathrm{d}\tilde{r}_{f_j, B_n} \, \mathrm{d}\tilde{r}_{f_n, B_n} \, \mathrm{d}t, \tag{8.20}$$

where $\mathcal{M}_{I_n}(t)$ and $\mathcal{M}_{S_n, I_n}(t) = \mathcal{M}_{S_n}(t)\mathcal{M}_{I_n}(t)$ denote the MGF of the macro-cell interference and joint MGF of the received signal and interference, respectively. In particular, $\mathcal{M}_{I_n}(t)$ and $\mathcal{M}_{S_n}(t)$ are defined as $\mathcal{M}_{I_n}(t) = \mathbb{E}[e^{-t \sum_{j=1, j \neq n}^{N} P_{f_j, B_n}^{\text{rx}}(\tilde{r}_{f_j, B_n})}]$ and $\mathcal{M}_{S_n}(t) = \mathbb{E}[e^{-t P_{f_n, B_n}^{\text{rx}}(\tilde{r}_{f_n, B_n})}]$, respectively. Because of the computational complexity of (8.19) and (8.20), this section focuses on finding analytical bounds on the capacity, and thereby bounds on the ASE of two-tier HetNets. It is important to note that determining the statistics of SINR for the two-slope path-loss model is computationally intensive mainly due to the arbitrary locations of interferers.

8.5.2 Assumptions to Derive Upper and Lower Bounds

This section first lists the assumptions used to derive the analytical bounds for the ASE of the COE deployment. Nonetheless, such constraints have been relaxed in simulations to provide a fair comparison. Next, a well-known established method for computing upper and lower bounds is utilized by fixing the distance of the macro-cell and small-cell interferers [233]. Since a location-based power control is employed, fixing the distance of interferers ultimately leads to fixing the transmission power of the interferers. Nonetheless, the second assumption of fixed transmission powers is not considered deliberately, rather it is a consequence of the previously considered assumption of fixed distance of the interferers. The worst (near) and best (far) location of the interferers refers to the upper and lower bounds, respectively.

Figure 8.6 illustrates the geometrical model of the uplink interference in both macro-cell and small-cell networks showing the worst and best case distances of interferers to derive lower and upper bounds for the ASE of HetNet.

The mobile users in the small-cell networks adapt their transmission power according to (8.7), which is significantly less than the maximum transmitting power of the mobile

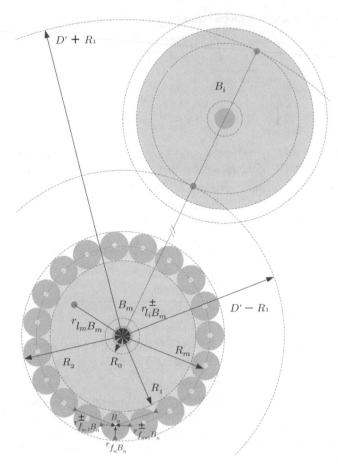

Figure 8.6 Geometrical illustration of uplink interference showing the worst- and best-case distance of the interferers in both macro and small cellular networks

users $P_{\max}$. The magnitude of the uplink interference signal received at the small-cell BS depends significantly on the transmission powers of the small-cell mobile users (or more explicitly interferers), which, in turn, depend on their battery power, target signal quality level and employed power control scheme. For example, the transmission power of a small-cell mobile user with $R_n = 50\,\text{m}$, $P_{\max} = 1\,\text{W}$ and $P_0 = 0.8\,\mu\text{W}$ is anticipated to be less than $0.5\,\text{mW}$ by using (8.7). Due to such low uplink transmission powers, the interference level received from the interfering $N - 1$ small-cells, particularly those deployed far away from the reference small-cell is significantly weak and can be considered negligible. On the basis of this reason, the interference only from adjacent small-cells is considered to derive the bounds.[6]

---

[6] In [234], the output power of the mobile phones was measured by Swedish TeliaSonera 3G Networks and it was concluded that the transmission power of the mobile phone depends on the type of the environment, that is, rural, urban, indoor or outdoor, and the type of application, that is, voice or video. Their measurements showed that the mobile phones in the current and future generation networks transmit in the power range of 0.2–0.5 mW, depending on the environment and application.

At this point, it is important to stress further that the considered assumption of the interference coming from only the adjacent small-cells instead of the $N - 1$ interfering small-cells is significantly useful in improving the analytical tractability while providing clean closed-form bounds for the ASE of COE configuration. Regarding the macro-cell network, it is considered that the interfering signal is received from the worst and best interfering mobile users in each of the $M - 1 = 6$ macro-cells. This simplification facilitates evaluating the bounds for the worst and best case interference scenarios in the most efficient manner.

- **Upper Bound:** Consider a macro-cell best interference configuration which corresponds to the case where all co-channel interferers are located on the far boundary of their respective cells, that is, at a distance given by

$$d_m^+ = D' + (R_m - R_n) \tag{8.21}$$

from the desired mobile BS and transmitting with power $P_m = P_{\max}$. Similarly, the small-cell interferers are considered to be located at a distance given by

$$d_n^+ = 2R_n + R_n. \tag{8.22}$$

The interferers are considered to transmit with the fixed transmitting powers applicable at the cell edge of small-cell, $P_n = \min(P_{\max}, P_0 R_n^{\alpha_{1n}}(1 + R_n/g_n)^{\beta_n})$. The desired small-cell mobile user is considered to transmit with an *adaptive* transmission power.
- **Lower Bound:** Consider a macro-cell worst interference configuration which corresponds to the case where all the co-channel interferers are located near the boundaries of their respective cells, that is, at a distance given by

$$d_m^- = D' - (R_m - R_n) \tag{8.23}$$

from the desired mobile BS and transmit with power $P_m = P_{\max}$. Similarly, small-cell interferers are considered to be located at a distance

$$d_n^- = 2R_n - R_n. \tag{8.24}$$

The interferers transmit with fixed transmitting powers applicable at the cell edge of the small-cell, $P_n = \min(P_{\max}, P_0 R_n^{\alpha_{1n}}(1 + R_n/g_n)^{\beta_n})$. The desired small-cell mobile user is considered to transmit with an *adaptive* transmission power.

8.5.3 Analytical Bounds on the Capacity of Macro-cell Network

By assuming the worst and best interfering mobile users in a macro-cell network, the SINR of the macro-cell mobile user can be evaluated by substituting (8.21) and (8.23) into (8.14) as follows:

$$\gamma_{l_m}^{\pm} = \frac{S_m}{I_m^{\pm} + \sigma^2} = \frac{P_{\max}\zeta_{l_m,B_m} r_{l_m,B_m}^{-\alpha_{1m}}(g_m + r_{l_m,B_m})^{-\beta_m}}{\sum_{i=1,i\neq m}^{M} P_{\max}\zeta_{l_i,B_m}(d_m^{\pm})^{-\alpha_{1m}}(g_m + d_m^{\pm})^{-\beta_m} + \sigma^2}, \tag{8.25}$$

where S_m denotes the desired signal power and $I_m^{\pm}$ represents the bounded cumulative interference received at the BS of interest of macro-cells. The bounds for (8.19) can be expressed as

$$C_{l_m}^{\pm} = w_{l_m} \int_{R_0}^{R_1} \hat{C}_{l_m}^{\pm}(r) f_{r_{l_m,B_m}}(r) dr_{l_m,B_m} \tag{8.26}$$

$$= w_{l_m} \int_0^{\infty} \int_{R_0}^{R_1} \frac{\mathcal{M}_{I_m}^{\pm}(t) - \mathcal{M}_{S_m,I_m}^{\pm}(t)}{t} e^{-\sigma^2 t} f_{r_{l_m,B_m}}(r) dr_{l_m,B_m} dt, \tag{8.27}$$

where $\mathcal{M}_{I_m}^{\pm}(t)$ denotes the lower and upper bounds for the MGF of the bounded cumulative interference received at the BS of interest and $\mathcal{M}_{S_m}(t)$ denotes the exact MGF of the desired signal in the macro-cell networks. By assuming i.i.d interfering mobile users in $M-1$ interfering macro-cells, the MGF of the bounded cumulative interference $I_m^{\pm}$ can be evaluated as follows:

$$\mathcal{M}_{I_m}^{\pm}(t) = \prod_{i=1,i\neq m}^{M} \mathcal{M}_{\zeta_{l_i,B_m}} \left(\frac{P_{\max} g_m^{\beta_m}(g_m + d_m^{\pm})^{-\beta_m}}{(d_m^{\pm})^{\alpha_{1m}}} t \right)$$

$$= \left(\mathcal{M}_{\zeta_{l_i,B_m}} \left(\frac{P_{\max} g_m^{\beta_m}(g_m + d_m^{\pm})^{-\beta_m}}{(d_m^{\pm})^{\alpha_{1m}}} t \right) \right)^{M-1}. \tag{8.28}$$

Similarly, the MGF of the received signal, $\mathcal{M}_{S_m}(t)$ can be derived by the use of the scaling property of MGF as follows:

$$\mathcal{M}_{S_m}(t) = \mathcal{M}_{\zeta_{l_m,B_m}} \left(\frac{P_{\max}(1 + r_{l_m,B_m}/g_m)^{-\beta_m}}{r_{l_m,B_m}^{\alpha_{1m}}} t \right). \tag{8.29}$$

Moreover, the joint MGF is given by $\mathcal{M}_{S_m,I_m}^{\pm}(t) = \mathcal{M}_{S_m}(t)\mathcal{M}_{I_m}^{\pm}(t)$.

The expression in (8.26) represents the generalized bounds on the mean achievable capacity of the desired mobile user in macro-cell networks over any type of fading channel that assumes knowledge of the MGF of the composite fading distribution.

8.5.4 Analytical Bounds on the Capacity of Small-Cell Networks

By assuming the worst and best interfering mobile users in a small-cell network, the SINR of the macro-cell mobile user can be derived by substituting (8.22) and (8.24) into (8.18) as follows:

$$\gamma_{f_n}^{\pm} = \frac{S_n}{I_n^{\pm} + \sigma^2} = \frac{P_{f_n,B_n}^{tx}(\tilde{r})\zeta_{f_n,B_n}\tilde{r}_{f_n,B_n}^{-\alpha_{1n}}(g_n + \tilde{r}_{f_n,B_n})^{-\beta_n}}{\sum\limits_{j=n\pm1} P_n\zeta_{f_j,B_n}(d_n^{\pm})^{-\alpha_{1n}}(g_n + d_n^{\pm})^{-\beta_n} + \sigma^2}, \tag{8.30}$$

where S_n denotes the desired signal power and $I_n^{\pm}$ denotes the bounded cumulative interference received at the BS of interest of a small-cell. The bounds on (8.20) can be derived as follows:

$$C_{f_n}^{\pm} = w_{f_n} \int_0^{R_n} \hat{C}_{f_n}^{\pm}(\tilde{r}) f_{\tilde{r}_{f_n,B_n}}(\tilde{r}) d\tilde{r}_{f_n,B_n} \tag{8.31}$$

$$= w_{f_n} \int_0^{\infty} \int_0^{R_n} \frac{\mathcal{M}_{I_n}^{\pm}(t) - \mathcal{M}_{S_n,I_n}(t)^{\pm}}{t} e^{-\sigma^2 t} f_{\tilde{r}_{f_n,B_n}}(\tilde{r}) d\tilde{r}_{f_n,B_n} dt, \tag{8.32}$$

where $\mathcal{M}_{I_n}^{\pm}(t)$ denotes the lower and upper bounds on the MGF of the bounded cumulative interference received at the BS of interest and $\mathcal{M}_{S_n}(t)$ denotes the exact MGF of the desired signal in small-cell networks. As stated earlier, the location of the worst and best interferers in small-cell networks is shown in Figure 8.6. In general, $\mathcal{M}_{\tilde{I}_n}^{\pm}(t)$ can be expressed by using the scaling property of MGF as follows:

$$\mathcal{M}_{\tilde{I}_n}^{\pm}(t) = \prod_{j=n\pm1} \mathcal{M}_{\zeta_{f_j},B_n} \left(P_n (2R_n \pm R_n)^{-\alpha_{1n}} \left(1 + \frac{2R_n \pm R_n}{g_n} \right)^{-\beta_n} t \right). \tag{8.33}$$

Similarly, the MGF of the desired signal, $\mathcal{M}_{S_n}(t)$ can be expressed as

$$\mathcal{M}_{S_n}(t) = \mathcal{M}_{\zeta_{f_n},B_n} \left(\frac{P_{f_n,B_n}(\tilde{r})(1 + \tilde{r}_{f_n,B_n}/g_n)^{-\beta_n}}{\tilde{r}_{f_n,B_n}^{\alpha_{1n}}} t \right). \tag{8.34}$$

The expressions in (8.31) represent the generalized bounds on the mean achievable capacity of the desired mobile user in small-cell networks over any type of fading channel that assumes knowledge of the MGF of the composite fading distribution.

8.6   Analytical Bounds on ASE over Generalized-$\mathcal{K}$ Fading Channel

The CDF and MGF of the generalized-$\mathcal{K}$ distribution involves Meijer-G and Whittaker functions, respectively, which reduce the analytical tractability because of their computational complexity. However, in order to avoid the associated computational difficulties, the authors in [229] proposed an accurate approximation of the generalized-$\mathcal{K}$ distribution by a more tractable gamma distribution using the moment-matching method, that is, $\mathcal{K}_G(m_c, m_s, \Omega) \approx$ Gamma$(m_{(\cdot)}, \theta_{(\cdot)})$. By matching the first and second moments of the two distributions, the corresponding values of $m_{(\cdot)}$ and $\theta_{(\cdot)}$ are given by Al-Ahmadi and Yanikomeroglu [229]

$$m_{(\cdot)} = \frac{m_c m_s}{m_c + m_s + 1 - m_c m_s \epsilon}, \quad \theta_{(\cdot)} = \frac{\Omega}{m_{(\cdot)}}, \tag{8.35}$$

where ϵ represents the adjustment factor. Let m_d and θ_d denote the fading severity (shape) and scale parameter, respectively, for the desired mobile users, which are located in both the macro-cell and small-cell networks. Similarly, let m_i and θ_i denote the fading severity (shape) and scale parameter, respectively, for the interfering mobile users, located in both the macro-cell and small-cell networks.

The analytical bound of (8.19) over a generalized-$\mathcal{K}$ fading channel is evaluated by using (8.26), where the MGFs of the bounded cumulative interference $\mathcal{M}_{\tilde{I}_m}^{\pm}(t)$ and the desired signal $\mathcal{M}_{S_m}(t)$ in the macro-cell network are determined by using (8.28) and (8.29), respectively:

$$\mathcal{M}_{\tilde{I}_m}^{\pm}(t) = \left(1 - \frac{g_m^{\beta_m}(g_m + D' \pm R_1)^{-\beta_m} \theta_i}{(D' \pm R_1)^{\alpha_{1m}}} t \right)^{-m_i(M-1)}, \tag{8.36a}$$

$$\mathcal{M}_{S_m}(t) = \left(1 - \frac{(1 + r_{l_m,B_m}/g_m)^{-\beta_m} \theta_d}{r_{l_m,B_m}^{\alpha_{1m}}} t \right)^{-m_d}. \tag{8.36b}$$

Similarly, the analytical bound of (8.20) over the generalized-$\mathcal{K}$ fading channel is determined by using (8.31), where the MGFs of the bounded cumulative interference $\mathcal{M}_{\tilde{I}_n}^{\pm}(t)$ and the desired signal $\mathcal{M}_{S_n}(t)$ in the small-cell network are derived by using (8.33) and (8.34), respectively, as follows:

$$\mathcal{M}_{I_n}^{\pm}(t) = \left(1 - P_n(2R_n \pm R_n)^{-\alpha_{1n}}\left(1 + \frac{2R_n \pm R_n}{g_n}\right)^{-\beta_n}\theta_i t\right)^{-m_i(2)}, \qquad (8.37a)$$

$$\mathcal{M}_{S_n}(t) = \left(1 - \frac{P_{f_n,B_n}(\tilde{r})(1 + \tilde{r}_{f_n,B_n}/g_n)^{-\beta_n}\theta_d}{\tilde{r}_{f_n,B_n}^{\alpha_{1n}}}t\right)^{-m_d}. \qquad (8.37b)$$

The desired and interference signals in both the macro-cell and small-cell networks assume a gamma distribution with different shape and scale parameters. Therefore, in order to derive a closed-form expression for the conditioned capacity, next a general result is derived, which is applicable to both macro-cells and small-cells.

Theorem 8.1 (Closed-form expression for the conditioned capacity in the interference-limited regime) *Consider two gamma random variables X and Y such that $X \sim \mathrm{Gamma}(a,b)$ and $Y \sim \mathrm{Gamma}(c,d)$. A closed-form expression for the conditional bounded capacity $\hat{C}_{(\cdot)}^{\pm}(\cdot)$ as defined in (8.26) and (8.31) for the macro-cell and small-cell mobile user, respectively, is given by*

$$\hat{C}_{(\cdot)}^{\pm}(\cdot) = \int_0^{\infty} \frac{(1+at)^{-b}}{t}dt - \frac{(1+at)^{-b}(1+ct)^{-d}}{t}dt$$

$$= \sum_{k=1}^{d} \binom{d}{k} \mathbf{B}(k, b+d-k) \,_2F_1\left[b, k, b+d, 1 - \frac{a}{c}\right], \qquad (8.38)$$

where $\mathbf{B}(p,q) = (\Gamma(p)\Gamma(q)/\Gamma(p+q))$ denotes the beta function; $_2F_1[(\cdot),(\cdot);(\cdot);(\cdot)]$ denotes the Gauss Hypergeometric function, a, b, c and $d \in \mathbb{R}^+$ are scaling constants and d and k are integers.

Proof. See Appendix B.

Applying the result derived in Theorem 8.1 and using (8.6) and (8.6), $\hat{C}_{l_m}^{\pm}(r)$ and $\hat{C}_{f_n}^{\pm}(\tilde{r})$ can be expressed, respectively, as

$$\hat{C}_{l_m}^{\pm}(r) = \sum_{k=1}^{m_d} \binom{m_d}{k} \mathbf{B}(k, m_i(M-1) + m_d - k) \,_2F_1[m_i(M-1), k,$$

$$m_i(M-1) + m_d, 1 - \frac{r_{l_m,B_m}^{\alpha_{1m}}(g_m + r_{l_m,B_m})^{\beta_m}\theta_i}{(D' \pm R_1)^{\alpha_{1m}}(g_m + D' \pm R_1)^{\beta_m}\theta_d}\right] \qquad (8.39)$$

and

$$\hat{C}_{f_n}^{\pm}(\tilde{r}) = \sum_{k=1}^{m_d} \binom{m_d}{k} \mathbf{B}(k, 2m_i + m_d - k) \,_2F_1[2m_i, k,$$

$$2m_i + m_d, 1 - \frac{P_n(2R_n \pm R_n)^{-\alpha_{1n}}(g_n + 2R_n \pm R_n)^{-\beta_n}\theta_i}{P_{f_n,B_n}(\tilde{r})\tilde{r}_{f_n,B_n}^{-\alpha_{1n}}(g_n + \tilde{r}_{f_n,B_n})^{-\beta_n}\theta_d}\right] \qquad (8.40)$$

By substituting (8.39) into (8.26) and (8.40) into (8.31), the bounds on the mean achievable capacity of the desired mobile user in the macro-cell and small-cell networks are obtained

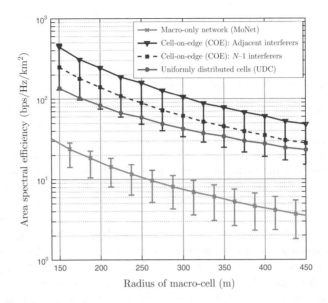

Figure 8.7   Analytical bounds on the ASE of (i) COE configuration considering that the interferers are located at the worst and best distances in each of the two adjacent small-cells and M co-channel macro cells and (ii) MoNet configuration as a function of the radius of the macro-cell

under the interference-limited regime. Also, the bounds for (8.9) for a COE configuration is given by

$$\eta^{\pm} = \frac{4\mathcal{C}_h^{\pm}}{\pi w_h R_u^2 (R_m + R_n)^2},\tag{8.41}$$

where $\mathcal{C}_h^{\pm} = \mathcal{C}_m^{\pm} + \mathcal{C}_n^{\pm}, \mathcal{C}_m^{\pm} = \sum_{l=1}^{U_m} \mathcal{C}_{l_m}^{\pm}$ and $\mathcal{C}_n^{\pm} = \sum_{n=1}^{N} \sum_{f=1}^{U_f} \mathcal{C}_{f_n}^{\pm}$.

Figure 8.7 illustrates the best and worst bounds on the ASE of MoNet and COE configuration. The bounds provide insights into the gain and loss in the ASE of the desired mobile user in the best and worst case interference conditions, respectively. It is observed that the analytical upper bound on the ASE of the COE configuration is quite tight in the presence of interferers from adjacent cells. Also, the analytical bounds are observed to be useful in capturing the ASE of COE and UDC configurations with $N-1$ interferers. The lower bound is comparatively loose. However, it illustrates the worst-case ASE when the macro-cell and small-cell interferer's location is in the neighbourhood of the desired cell centre. Despite the ASE degradation, the achieved ASE is higher than that corresponding to MoNet.

8.7   Energy Analysis of HetNets

This section quantifies the energy improvements of HetNets in terms of energy consumption, energy savings and associated energy economics. The mapping between the power consumption/savings to energy consumption/savings can be understood from the following relationship:

$$\text{Energy} = \text{Power} \cdot \phi(t) \cdot \text{No. of Days/year} \quad (\text{kWh/year}),\tag{8.42}$$

where $\phi(t)$ denotes the number of hours per day a mobile user is active under full-load conditions.

8.7.1 Energy Consumption of Two-Tier HetNets

In general, energy consumption is defined as the power consumption per unit time such that the uplink power consumption can be directly calculated using (8.7) and the associated energy consumption can be calculated using (8.42).

Figure 8.8a depicts the energy consumption per user for HetNets with COE deployment as a function of small-cell radius. It can be seen clearly that the energy consumption of the COE deployment outperforms the energy consumption of (i) UDC deployment and (ii) MoNets (compare the solid curve with dashed and dotted curves). The significant improvement is due to the fact that the small-cells around the edge of the macro-cell ensure a reduction in the number of edge mobile users of the macro-cell that transmit at their maximum power.[7] At this point, it is important to emphasize that MoNet is a state of the COE deployment when small-cells are inactive. The resultant coverage radius of the macro-cell is $R_m + R_n$ given the geometrical illustration shown in Figure 8.1. Therefore, with the increase in the small-cell radius R_n more mobile users will be located around the edge of the cell, and will transmit with the maximum power. This is the primary reason of increase in energy consumption for MoNet when the small-cells are inactive. The same reason applies to Figure 8.8b as well. The comparative summary on the performance of HetNets with COE deployment with respect to the two competitive network deployments is next presented.

- Comparison with MoNets: Energy consumption of the COE deployment outperforms the energy consumption of the MoNets due to (i) the deployment of the small-cells and (ii) reduction in the number of cell-edge mobile users who transmit with their maximum power. As an example, for $R_n = 50\,\text{m}$, the energy consumption of the COE deployment reduces to 1 kWh per user, which offers a 68% reduction in energy consumption compared with MoNets.
- Comparison with the UDC deployment: Energy consumption of the COE deployment outperforms the UDC deployment mainly due to the reduction in the cell-edge mobile users who transmit with their maximum power, for example, for $R_n = 50\,\text{m}$, the COE deployment offers a 37% reduction in energy consumption compared with the UDC deployment.

8.7.2 Energy Savings of Two-Tier HetNets

Power savings per mobile user is assessed by using (8.7) as $P_{\max} - P_0 \frac{r^\alpha (1+r/g)^\beta}{K}$. The associated energy savings is calculated by using the relationship introduced in (8.42).

Figure 8.8b depicts the amount of energy saved by the mobile users who transmit with an adaptive power, for example, the energy savings offered by the COE deployment at $R_n = 100\,\text{m}$ is 4 kWh, which is more than double the savings that the network can achieve

---

[7] In [235], a threshold distance, R_t was calculated and it is referred to as the distance beyond which the mobile users are required to transmit with the maximum power. As an example, with $\alpha = \beta$, $R_t = 422\,\text{m}$ such that the number of mobile users transmitting with the maximum power increases with the increase in R_m beyond R_t.

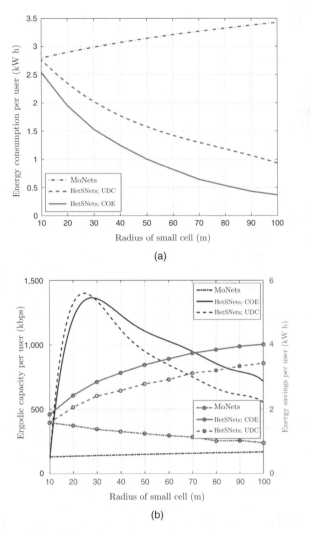

Figure 8.8 Summary of energy analysis per user as a function of small-cell radius. (a) Energy consumption; (b) spectral and energy gains

at $R_n = 10$ m and which is 1.9 kWh. In addition, Figure 8.8b quantifies the average capacity achieved per user as a function of R_n. It can be observed that the HetNets with the COE deployment remain spectrally efficient over medium to high range of values for R_n (for more detailed results, discussions and mathematical interpretations, see [235]).

8.8   Ecology and Economics of HetNets

This section presents the ecological impact of energy consumption and energy savings of the HetNets in terms of CO_2e emissions and the associated economics of the networks.

8.8.1 CO_2e Emissions and Reduction in CO_2e Emissions

In order to determine the ecological impact of the energy consumption of HetNets, this section calculates the corresponding CO_2e emissions in mega tonnes [Mtonnes]. The conversion factor used to convert the energy consumption into CO_2e emissions is $1\,kWh = 0.5246\,kg\ CO_2e$ emissions, and it represents the energy used at the point of final consumption [236].

Figure 8.9a illustrates the uplink CO_2e emissions for (i) MoNets; (ii) HetNets with UDC deployment and (iii) HetNets with COE deployment, where all mobile users are transmitting with their adaptive power to maintain the desired SINR of the link. The CO_2e emissions of the systems under consideration are compared with the CO_2e emissions of the MoNets without PC, that is, the network where the mobile users are transmitting with the maximum power and the small-cells are inactive. It can be seen clearly that the CO_2e emissions of the Het-Nets are reduced significantly in comparison with the MoNets without PC. As an example, the CO_2e emissions of the MoNets without PC in 2016 is approximately 19 Mtonnes. The MoN-ets with PC reduce the estimated CO_2e emissions to 13 Mtonnes (30% reduction). This can be further reduced to 8 Mtonnes (67% reduction) by introducing small-cells in HetNet with COE deployment. Finally, the significant reduction in CO_2e emissions can be achieved by introducing small-cells around the edge of the macro-cells. The proposed HetNets with COE deployment guarantees the reduction of the CO_2e emission to 3.5 Mtonnes (82% reduction). Therefore, the mobile communications industry can enforce effective policies to reduce the global carbon footprint emissions.

8.8.2   Daily CO_2e Emissions Profile

The daily CO_2e emissions profile quantifies the amount of CO_2e emissions corresponding to the various mobile traffic loads, that is, percentage of the active mobile users at different times of the day. Figure 8.9b depicts the daily CO_2e emissions profile of an European country corresponding to the daily mobile traffic loads profile presented in [60]. It can be seen clearly that the CO_2e emissions of MoNets without PC are significantly higher during peak times of the day. Moreover, the CO_2e emissions of the HetNets with COE deployment improve significantly during the peak periods of the day compared with the other two competitive network deployments (MoNets with PC and HetNets with UDC deployment). As an example, the maximum number of active users is 16% at 9 pm. The corresponding daily CO_2e emissions of MoN-ets without PC is estimated as 142 Mtonnes, and it decreases to 120 Mtonnes in the presence of PC. Moreover, the UDC deployment contributes 60 Mtonnes to daily CO_2e emissions. In addition, HetNet with COE deployment reduces the daily CO_2e emissions to 47.5 Mtonnes. Therefore, the daily CO_2e emissions profile clearly shows that the proposed HetNets with COE deployment improves the energy savings, and thereby they establish green HetNets by contributing less amounts of CO_2e emissions to the environment.

8.8.3   Low-Carbon Economy

The world economy has witnessed three economic transformations: (1) the industrial revolution, (2) the technological revolution and (3) the modern era of globalization. At present, the world economy stands at the edge of the next transformation: the age of green economy. The

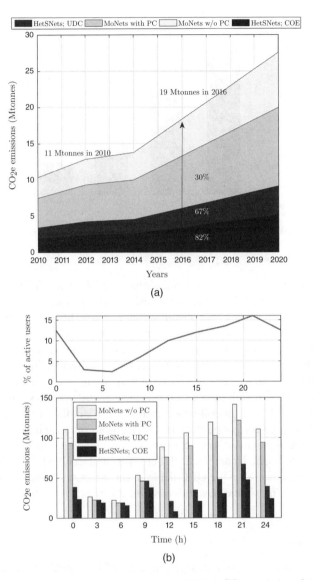

Figure 8.9 Summary of carbon footprint of HetNets. (a) Uplink CO_2e emissions for several networks; (b) Daily CO_2e emissions profile corresponding to various traffic loads

green economy is an economic development based on ecological sustainability and knowledgeable decisions. Most of the developing and emerging economies are struggling to balance the economical and environmental resources, at both local and global scales. The ICT and mobile communication industries are required to act now and contribute toward mitigating the effects of climate change and reducing the global carbon footprint.

The low-carbon economy index (LCEI) is generally defined as the amount of CO_2 emissions released per capita gross domestic product (GDP) and is fundamentally dependent on several

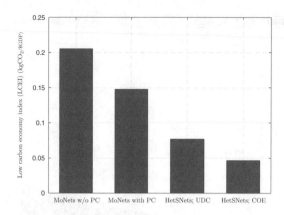

Figure 8.10   Low carbon economy index (LCEI) for several competitive network configurations

factors including energy efficiency, CO_2 emissions, population density and economic infrastructure. In particularly, the LCEI of a mobile user is the measure of CO_2 emissions corresponding to the energy consumed over the uplink per capita GDP. Figure 8.10 shows the LCEI
of a mobile user under several competitive network configurations, namely (i) macro-only
networks without PC, (ii) macro-only networks with PC, (iii) HetNets with COE and (iv) Het-
Nets with UDC. Here, per capita GDP is assumed as $ 12,000 as mentioned in the World Bank
statistics [237]. It can be seen clearly that LCEI of the heterogeneous networks can be reduced
significantly in comparison with the LCEI of the macro-only networks. The improvement in
LCEI is due to the fact that the mobile users in HetNets adapt their transmission power, and
thereby reduce the energy consumption and CO_2 emissions of the uplink. However, the LCEI
of HetNets with COE deployment is much less than other competitive networks, including Het-
Nets with UDC deployment. Under the COE deployment, the cell-edge mobile users transmit
with a much reduced power than the UDC deployment, where a significant number of mobile
users transmit with the maximum power to meet the desired target SINR.

8.9   Summary

In this chapter, we discussed the uplink performance of two-tier HetNets, where small-cells
are arranged at the edge of the macro-cell, that is, the COE configuration, which is shown to
facilitate the cell-edge mobile users with a guaranteed high-quality link, and thereby it tends
to increase the ASE compared with the other two competitive configurations, namely the UDC
and MoNet configurations. The channel propagation model explicitly considers the strong LOS
conditions that exist mainly in the small-cell scenario. Analytical bounds are derived to illus-
trate the ASE of HetNets under the worst and best interference scenarios. The bounds are
generalized for any composite fading distribution and closed-form expressions are presented
for generalized-$\mathcal{K}$ fading channels. It is shown that significant energy savings can be achieved
by (i) deploying small-cells around the edge of macro-cells and (ii) employing PC in the uplink
where each mobile user transmits with adaptive power. It is shown further that the CO_2e emissions of the COE deployment is reduced to 82% in comparison with the CO_2e emissions of

the MoNets without employing PC. Therefore, the reduction in CO_2e emissions is considered as a cornerstone in designing and planning environment-friendly wireless networks.

APPENDIX A - Simulation Parameters

Table 8.1 Simulation parameters for COE deployment in HetNet

Simulation parameter	Small-cell	Macro-cell
Transmission power (P_{max})	1 W	1 W
Cell radius ($R_{(\cdot)}$)	50 m	150–500 m
Path-loss exponent ($\alpha_{1(\cdot)}$)	1.8	2.0
Path-loss exponent ($\alpha_{2(\cdot)}$)	3.6	4.0
Additional path-loss exponent ($\beta_{(\cdot)}$)	1.8	2.0
BS antenna height (h_{rx})	12.5 m	25 m
Mobile antenna height (h_{tx})	2 m	2 m
Reference distance (R_0)		1 m
Target power received (P_0)		0.008 mW
Breakpoint distance ($g_{(\cdot)}$)	1,300 m	500 m
System bandwidth (w_t)	20 MHz	
Reuse factor (R_u)	2	
Small-cell population factor (CPF)	1	
Thermal noise power (σ^2)	2×10^{-21} W/Hz	
Macro-cell user density	0.005 m^{-2}	

APPENDIX B - Proof of (8.38)

Considering the integral representation of the bounded capacity conditioned on the location of the desired user (8.38) in the presence of a gamma composite fading channels, one can infer that

$$\hat{C}^{\pm}_{(\cdot)}(\cdot) = \int_0^\infty \frac{1}{t(1+at)^b} dt - \frac{1}{t(1+at)^b(1+ct)^d} dt, \tag{8.43}$$

and simple algebraic manipulations show that (8.43) can be re-written as follows:

$$\hat{C}^{\pm}_{(\cdot)}(\cdot) = \int_0^\infty \frac{(1+ct)^d - 1}{t(1+at)^b(1+ct)^d} dt. \tag{8.44}$$

Applying the binomial expansion formula in the factor $(1+ct)^d$ present in the numerator of (8.44), it follows that

$$(1+ct)^d = \sum_{k=0}^d \binom{n}{k} (ct)^k. \tag{8.45}$$

Substituting the value from (8.45) into (8.44), the integral in (1.44) can be written as follows:

$$\hat{C}_{(\cdot)}^{\pm}(\cdot) = \int_0^\infty \frac{\sum\limits_{k=1}^{d} \binom{n}{k}(tc)^k}{t(1+at)^b(1+ct)^d}dt, \tag{8.46}$$

and (8.46) can be further simplified to

$$\hat{C}_{(\cdot)}^{\pm}(\cdot) = \sum_{k=1}^{d}\binom{n}{k}\int_0^\infty \frac{(tc)^k}{t(1+at)^b(1+ct)^d}dt,$$

$$= \sum_{k=1}^{d}\binom{n}{k}\int_0^\infty \frac{t^{k-1}c^k}{(1+at)^b(1+ct)^d}dt. \tag{8.47}$$

Now, by using the identity [238][3.197/1], that is, $\int_0^\infty x^{\nu-1}(\beta+x)^{-\mu}(\gamma+x)^{-\varrho}dx = \beta^{-\mu}\gamma^{\nu-\varrho}\mathbf{B}(\nu,\mu-\nu+\varrho)\,_2F_1(\mu,\nu,\mu+\varrho,1-(\gamma/\beta))$ and carrying out simple algebraic manipulations, the integral in (8.43) can be evaluated and simplified to (8.38).

9

D2D Communications in Hierarchical HetNets

The growth in mobile communication systems has led to a tremendous increase in the energy consumed by the mobile networks. Device-to-device (D2D) communications and small-cell networks are considered to be an integral part of the 5G communications due to the low power, low cost and ease of deployment of small-cell BSs (SBSs) and D2D communications. This chapter introduces a three-tier hierarchical HetNet by exploiting D2D communications in traditional HetNets. D2D communications are deployed within the HetNet, where closely located mobile users are engaged in direct communication without routing the traffic through the cellular access network. The proposed configuration mandates reduction of the interference levels in the resultant HetNet by reducing the transmitter–receiver distance and ensuring that the mobile users are transmitting with adaptive power subject to maintaining their desired link quality. The performance of the proposed network configuration is investigated by comparing the spectral and backhaul energy efficiency improvements in the hierarchical HetNet against traditional HetNets. Simulation results show that the proposed deployment achieves a significant reduction in total transmission power compared with the full small-cell deployment. It is shown that the proposed network deployment outperforms the network with full small-cell deployment, and thus it provides a greener alternative to the small-cell deployment.

9.1 Introduction

Capacity and coverage enhancement have been major goals of every wireless communication system. With the advent of mobile data services and smart devices, the capacity requirements have exploded in recent years, and the worldwide mobile traffic forecast is expected to reach more than 127 exabytes (EBs) for 2020 [239]. An increase of 1,000-fold in wireless traffic is expected in 2020 as compared to the 2010 figures as well as an expected number of 50 billion communication devices [240]. This sudden growth of the mobile traffic can be handled by capacity enhancement, which mainly comprises three techniques: spectral efficiency, spectral aggregation and network densification [241]. The spectral efficiency approach mainly targets

Green Heterogeneous Wireless Networks, First Edition. Muhammad Ismail, Muhammad Zeeshan Shakir, Khalid A. Qaraqe and Erchin Serpedin.
© 2016 John Wiley & Sons, Ltd. Published 2016 by John Wiley & Sons, Ltd.

interference-aware and cooperative communications, for example, coordinated beamforming, multiple-input multiple-output (MIMO), coordinated multipoint (CoMP) and device-to-device cooperation. The spectrum aggregation consists of carrier aggregation to enhance the system bandwidth.

The network densification is globally accepted as the quick and cost-effective solution to meet capacity and coverage demands. The deployment of a huge number of small-cells was reported in the past [242], and it results in heterogeneous networks, where several types of low-power SBSs such as femto cell, pico cell and relays are deployed within a macro-cell BS (MBS) coverage area to improve the spectral efficiency and coverage of cellular networks. SBS deployments ensure better transmission quality due to the short distance between the small-cell users and the associated SBSs, and therefore, they improve the network spectral efficiency (SE) [243, 244]. It has been shown in [245] that the deployment of pico cells can improve the user throughput and expands the range of cells. In [246], the authors proposed an efficient distribution of femto cells within MBS based on the minimum allowable received signal power at the user. It was shown that the cell coverage area was increased twofold via efficient femto-cell location deployment. The authors in [247] proposed a heterogeneous deployment of femto cells around the cell edge of a macro cell to improve the area spectral efficiency (ASE) of the network. On the contrary, the SBS deployment in HetNets requires substantial infrastructure where the cellular traffic route through the SBS even in the situation where the communicating devices are close to each other [215, 248]. Moreover, SBS deployment requires an additional link to backhaul the traffic to the core cellular network, which increases the capital and operational expenditures for the operators [249–251].

With the spectral performance of the wireless links approaching the theoretical limits in the present cellular wireless networks, researchers have been working in the framework of LTE-Advanced to further facilitate the communications among mobile users in a ubiquitous and cost-effective manner. One of the means to increase the achievable rate in cellular communications is through direct communication between closely located mobile users. This form of communication is referred to as device-to-device (D2D) communication [252, 253]. Mobile devices involved in D2D communication form a direct link with each other, without the need of routing traffic via the cellular access network, which leads to lower transmission power and end-to-end delay, as well as freeing network resources. The lower transmission powers manifest through reduced interference levels in the system and battery power savings, while the improved rate is achieved as a result of the low path loss between any pair of devices involved in D2D communication [254].

In this context, this chapter proposes a three-tier hierarchical HetNet, where D2D communication is introduced as tier 3 network within MBS (tier 1) and SBS (tier 2) to improve the SE of the considered HetNet such that a percentage of the mobile users engages in D2D communications in both higher tiers. D2D communication signalling could be carried out through either the macro-cell access network or Wi-Fi access points. This deployment setting is compared with the traditional HetNet in terms of capacity enhancement.

9.2   Modelling Hierarchical Heterogeneous Networks

This section describes the network architecture, spectrum partitioning and transmission model of a hierarchical HetNet.

9.2.1   Network Architecture

The hierarchical HetNet comprises the following tiers:

- Tier 1: Macro-cell users connected to the MBS.
- Tier 2: Small-cell users connected to the SBS.
- Tier 3: D2D users connected to MBS and SBSs.

The following subsections will present the assumptions underlying the user distribution in the macro-cell and small-cell networks and the integration of D2D communications in heterogeneous networks.

9.2.1.1   Macro-cell Network

The network shown in Figure 9.1 contains $U = \mu_m \pi (R_m^2 - R_0^2)$ users distributed inside the circular ring with radii R_m and R_0, where R_m denotes the macro-cell radius, R_0 denotes the minimum distance between a mobile user and MBS and μ_m represents the user density per m^2 in the coverage area of MBS. For the sake of simplicity, only one MBS in the top tier is shown. However, we assume $Q - 1 = 6$ interfering co-channel MBSs near the reference MBS.

Let $U_m = (1 - \eta)U$ independent PPP distributed MTs be connected to MBS and let η denote the percentage of users that are offloaded to SBSs. According to [255], wireless usage is shifting indoors where the majority of mobile traffic occurs, approximately 80% is indoor and nomadic, rather than truly mobile. In this chapter, we assume η to be 80% so that the remaining 20% of users are connected to MBS. Therefore, D2D communication in MBS and SBS is emerging as a possible solution to address such modern mobile traffic patterns in HetNets.

Let $U_m^{D2D} = U_m \zeta_m^{D2D}$ denote the number of MBS users involved in D2D communication, such that the distance between any two communicating D2D communication users is d (m), as shown in Figure 9.1. Moreover, the parameter $0 \leq \zeta_m^{D2D} \leq 1$ denotes the content exchange information and it describes the probability that the devices exploit the caching in MBS, share the content (peer-to-peer networking, single-/multiple-hop relaying, etc.) and establish direct link over the D2D protocol. The parameter ζ_m^{D2D} may be modelled probabilistically as representing the usage of caching in MBS. Under such a modelling set-up, $U_m^{cu} = U_m(1 - \zeta_m^{D2D})$ yields the number of MBS cellular users.

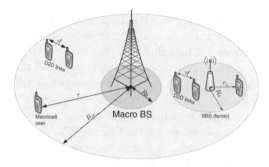

Figure 9.1   Hierarchical heterogeneous network showing MBS, SBS and D2D communication in the higher tiers

9.2.1.2 Small-Cell Network

Let $U_s = \pi \mu_s R_s^2$ denote the number of users in each SBS and μ_s define the user density of the sth SBS. The number of SBSs required to cover the MBS coverage area is

$$N = \left\lceil \frac{U - U_m}{U_s} \right\rceil, \tag{9.1}$$

where $\lceil x \rceil$ is the smallest integer not less than x and U_s denotes the number of users in the SBS.

Let ζ_s^{D2D} denote the content exchange information of the sth SBS, where the corresponding users are involved in D2D communication for device-centric and low-mobility indoor activities (gaming, ultrahigh-definition video sharing, etc). In this case, $U_s^{\text{cu}} = U_s(1 - \zeta_s^{\text{D2D}})$ yields the number of SBS cellular users (not involved in D2D communications), whereas the total number of users involved in D2D communication in the entire small-cell network can be expressed as

$$U_s^{\text{D2D}} = \pi R_s^2 \sum_{n=1}^{N} \mu_s \zeta_s^{\text{D2D}}, \tag{9.2}$$

where $U_s^{\text{D2D}}/2$ denotes the total number of D2D pairs in the small-cell network. The remaining users of all SBSs, not involved in D2D communication, are given by

$$U_s^{\text{tcu}} = U_s N - U_s^{\text{D2D}}. \tag{9.3}$$

9.2.2 D2D User Density in Hierarchical HetNets

The probability of users for D2D communication depends on many factors, including channel conditions and common contents. In order to choose D2D pairs in MBS (ζ_m^{D2D}) and SBS (ζ_s^{D2D}), the cumulative distribution function (CDF) for U_m^{D2D} MBS and U_s^{D2D} SBS users is approximated as shown in Figure 9.2.

Since MBS users are non-nomadic and fast moving as compared with SBS users; therefore, the value of $\zeta_m^{\text{D2D}} \sim 0.63$ is chosen and it shows approximately 50% probability for D2D users. For SBS users, the value of $\zeta_s^{\text{D2D}} \sim 0.70$ shows 60% probability for D2D users in a small-cell. For illustrative purposes, the value $\mu_m = 0.005$ for $R_m = 560$ m and $R_s = 30$ m is chosen, and it generates approximately 5,000 users among which 1,000 (20%) are MBS and the rest (80%) are SBS users. In Figure 9.2, the value of ζ_m^{D2D} corresponds to approximately 660 MBS D2D users, whereas ζ_s^{D2D} corresponds to approximately nine D2D users per SBS.

An ultra-dense environment is simulated by increasing the user density from 1 to 20 milli users/m^2. In order to deploy SBSs uniformly into the coverage area of MBS, the whole disc of radius R_m is divided into circular rings. For illustrative purposes, the rings of the hierarchical HetNet showing a three-tier network are shown in Figure 9.3.

In such a hierarchical network, the whole area is covered by the MBS with black circles showing SBS deployment. The small circles and plus signs show MBS and SBS users directly connected to the respective BSs. The crosses show D2D users either in MBS or SBSs as tier 3 network.

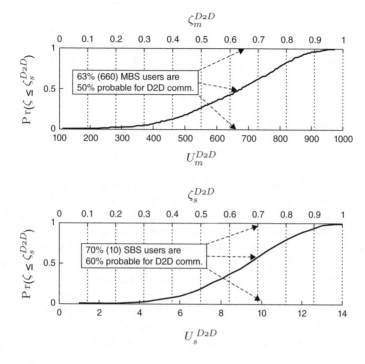

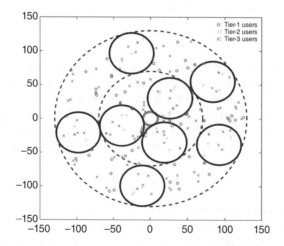

Figure 9.2 D2D user density based on the CDF approximation of ζ

Figure 9.3 Three-tier hierarchical HetNet showing only two-rings for illustrative purpose

9.2.3 Spectrum Partitioning in Hierarchical HetNets

We assume dedicated carrier deployment in the communication network, where the MBS, SBS and D2D communication users operate on separate bandwidths based on the active number of users associated with each technology. Let the total available spectrum be w_t (Hz). It follows that for the traditional HetNet,

$$w_t = w_m + w_s,$$

whereas for the hierarchical HetNet,

$$w_t = (w_m^{\mathrm{HH}} + w_m^d) + (w_s^{\mathrm{HH}} + w_s^d),$$

where $w_m = w_t(U_m/U)$ and $w_s = w_t(U_n N/U)$ are the dedicated channels of each MSB and SBS user in the traditional HetNet, respectively. Similarly, $w_m^{\mathrm{HH}} = w_t(U_m^{\mathrm{cu}}/U)$, $w_s^{\mathrm{HH}} = w_t(U_s^{\mathrm{tcu}}/U)$, $w_m^d = w_t(U_m^{\mathrm{D2D}}/U)$ and $w_s^d = w_t(U_s^{\mathrm{D2D}}/U)$ are the dedicated channels of each MBS, SBS, D2D with MBS and D2D with SBS user in the hierarchical HetNet, respectively. The number of channels in MBS and SBSs are assumed to be equal to the number of users they contain, and each channel is allocated to a single user [256]. Hence, the interference received at the MBS or SBS is from the mobile users in each of the neighbouring co-channel macro or small-cells that transmit on the same channel, while the interference in each D2D communication link is assumed to be from the closest D2D communication user that is not part of that communication link. This assumption was made because mobile devices engaged in D2D communication usually transmit with very low power, which causes reduced interference.

9.2.4 Power Control over D2D Links

The received signal power at a distance r between MBS or SBSs and one of the devices engaged in D2D communication is given by

$$P^{\mathrm{rx}} = P^{\mathrm{tx}} r^{-\alpha}(1 + r/g)^{-\beta}\,\Gamma, \tag{9.4}$$

where α and β denote the basic and additional path-loss exponents, respectively, and Γ denotes a path-loss-dependent constant. The parameter $g = 4H_{\mathrm{BS}}H_u/\lambda_c$ (m) is the break point of the path-loss curve, H_{BS} (m) represents the BS antenna height, H_u (m) denotes the mobile user antenna height and λ_c (m) denotes the wavelength of the carrier frequency F_c. Both the small-cell and D2D communication users are assumed to transmit with adaptive power while maintaining a certain received signal threshold. The adaptive transmission power of a user is given by

$$P_i^t = \min(P_i^{\mathrm{max}}, P_0 D_i^\alpha), \tag{9.5}$$

where $P_i^{\mathrm{max}}, P_0, D_i$ and α represent the maximum transmission power of a user, received signal power threshold, link distance and path-loss exponent, respectively. The assumption that $P_0 D_i^\alpha < P_i^{\mathrm{max}}$ is also considered due to the short link distances. This implies that

$$P_i^t = P_0 D_i^\alpha, \tag{9.6}$$

for all users in the network.

9.3 Spectral Efficiency Analysis

This section focuses on the spectral analysis of traditional and hierarchical HetNets.

9.3.1 Traditional HetNet

The sum rate of traditional HetNet (without D2D communication) consists of the individual sum rates of MBS and SBSs:

$$C = C_m + C_s = \sum_{l=1}^{U_m} C_{l,m} + \sum_{s=1}^{N} \sum_{z=1}^{U_s} C_{z,s}, \tag{9.7}$$

where C_m (bits/s) denotes the sum rate of MBS and C_s (bits/s) denotes the sum rate of SBSs. The achievable capacity $C_{l,m}$ of the lth user located in the mth MBS of a traditional HetNet is given by

$$C_{l,m} = w_m \, \mathbb{E}[\log_2(1 + \gamma_{l,m})]$$

$$= w_m \int_0^\infty \log_2(1 + \gamma_{l,m}) f_\gamma(\gamma_{l,m}) d\gamma_{l,m}, \tag{9.8}$$

where $f_\gamma(\gamma_{l,m})$ denotes the PDF of $\gamma_{l,m}$ and $\gamma_{l,m}$ is the signal-to-interference ratio (SIR) of the desired link. Assuming the thermal noise power is negligible compared with the co-channel interference power, the SIR of the lth user located in the mth macro cell is given by

$$\gamma_{l,m} = \frac{P_{l,m}^{rx}}{\sum_{\tau=1, \tau \neq m}^{Q} P_{l,\tau}^{rx}}, \tag{9.9}$$

where $P_{l,m}^{rx}$ (W) denotes the received power at the mth macro cell from the lth user and $\sum_{\tau=1, \tau \neq m}^{Q} P_{l,\tau}^{rx}$ denotes the sum of the individual interfering power levels received at the reference MBS from the interfering mobile users $\{l_\tau\}_{\tau=1, \tau \neq m}^{Q}$, which are located in each of the $Q - 1 = 6$ interfering MBSs. Substituting (9.4) into[1] (9.9), it turns out that the SIR of a macro-cell user is given by

$$\gamma_{l,m} = \frac{P_{l,m}^{tx} r_{l,m}^{-\alpha_m} (g_m + r_{l,m})^{-\beta_m}}{\sum_{\tau=1, \tau \neq m}^{Q} P_{l,\tau}^{tx} r_{l,\tau}^{-\alpha_m} (g_m + r_{l,\tau})^{-\beta_m}}. \tag{9.10}$$

Similarly $C_{z,s}$ is the achievable capacity of the zth user in the sth small-cell, and it is given by

$$C_{z,s} = w_s \, \mathbb{E}[\log_2(1 + \gamma_{z,s})]$$

$$= w_s \int_0^\infty \log_2(1 + \gamma_{z,s}) f_\gamma(\gamma_{z,s}) d\gamma_{z,s}, \tag{9.11}$$

---

[1] In (9.4), we suppress the notations for the sake of simplicity and better understanding.

where $f_\gamma(\gamma_{z,s})$ denotes the PDF of $\gamma_{z,s}$ and $\gamma_{z,s}$ denotes the SIR of the zth user in the sth small-cell, and is expressed as

$$\gamma_{z,s} = \frac{P_{z,s}^{\mathrm{rx}}}{\sum_{v=1,v\neq s}^{N} P_{z,v}^{\mathrm{rx}}}. \tag{9.12}$$

The parameter $P_{z,s}^{\mathrm{rx}}$ (W) in (9.12) represents the received powers at the sth small BS from the zth user and $\sum_{v=1,v\neq s}^{N} P_{z,v}^{\mathrm{rx}}$ is the sum of the power received at the sth small BS from the interfering small-cell users $\{z_v\}_{v=1,v\neq s}^{N}$ located in the neighbouring $N-1$ interfering small BSs in HetNet. Substituting (9.4) into (9.12), the SIR of the small-cell user is expressed as

$$\gamma_{z,s} = \frac{P_{z,s}^{\mathrm{tx}} r_{z,s}^{-\alpha_s}(g_s + r_{z,s})^{-\beta_s}}{\sum_{v=1,v\neq s}^{N} P_{z,v}^{\mathrm{tx}} r_{z,v}^{-\alpha_s}(g_s + r_{z,v})^{-\beta_s}}. \tag{9.13}$$

9.3.2 Hierarchical HetNet

The capacity of the hierarchical HetNet depends on the cellular and D2D users in both MBS and SBSs. In case of MBS, we have U_m^{cu} cellular and U_m^{D2D} D2D users, whereas for each SBS, we have U_s^{cu} cellular and U_s^{D2D} D2D users. The total capacity (bits/s) of the hierarchical HetNet is given by

$$\begin{aligned}C^{\mathrm{HH}} &= C_m^{\mathrm{HH}} + C_s^{\mathrm{HH}}\\ &= \sum_{l=1}^{U_m^{\mathrm{cu}}} C_{l,m}^{\mathrm{HH}} + \sum_{x=1}^{U_m^{\mathrm{D2D}}} C_{x,m}^{\mathrm{HH}} + \sum_{s=1}^{N}\left[\sum_{z=1}^{U_s^{\mathrm{cu}}} C_{z,s}^{\mathrm{HH}} + \sum_{x=1}^{U_s^{\mathrm{D2D}}} C_{x,s}^{\mathrm{HH}}\right],\end{aligned} \tag{9.14}$$

where C_m^{HH} consists of the capacity of U_m^{cu} cellular and U_m^{D2D} D2D users of MBS. Similarly, the capacity C_s^{HH} is the capacity of U_s^{cu} cellular and U_s^{D2D} D2D users of each SBS. Variables $C_{l,m}^{\mathrm{HH}}$ and $C_{z,s}^{\mathrm{HH}}$ represent the achievable capacity of MBS and SBS cellular users calculated similarly to (9.8) and (9.11), respectively.

The achievable capacity of the xth D2D communication user in MBS or SBS is expressed as

$$C_{x,y}^{\mathrm{HH}} = w_y^d \, \mathbb{E}[\log_2(1+\gamma_x)] = w_y^d \int_0^\infty \log_2(1+\gamma_x) f_\gamma(\gamma_x) \mathrm{d}\gamma_x, \tag{9.15}$$

for $y \in \{m,s\}$. Let $f_\gamma(\gamma_x)$ denote the PDF of the desired SIR γ_x of the xth D2D communication user in MBS or SBS. Then,

$$\gamma_x = \frac{P_{x,y}^{\mathrm{rx}}}{P_{x,i}^{\mathrm{rx}}}, \tag{9.16}$$

where $P_{x,y}^{\mathrm{rx}}$ denotes the xth D2D user's received power at its D2D partner in MBS or SBS and $P_{x,i}^{\mathrm{rx}}$ denotes the received interference power at the xth D2D user from the interfering D2D user i.

Substituting (9.4) into (9.16), the SIR of the xth mobile user ($y \in \{m,s\}$) is expressed as

$$\gamma_x = \frac{P_{x,y}^{\mathrm{tx}} d_{x,y}^{-\alpha_d}(g_d + d_{x,y})^{-\beta_d}}{P_{i,y}^{\mathrm{tx}} d_{i,y}^{-\alpha_d}(g_d + d_{i,y})^{-\beta_d}}. \tag{9.17}$$

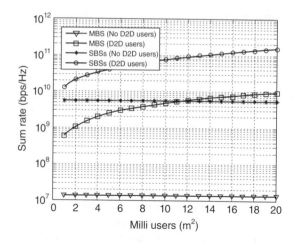

Figure 9.4   Sum Rate of MBS, SBSs with/without D2D users

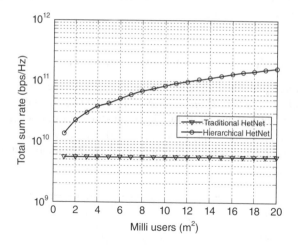

Figure 9.5   Total Sum Rate of HetNet and hierarchical HetNet

The capacity enhancement of the hierarchical HetNet is compared with the traditional HetNet in Figures 9.4 and 9.5. Figure 9.4 illustrates the sum rate (bps/Hz) versus the variable user density for MBS and SBSs for the two cases of non-D2D and D2D users. The sum rate capacity increases with an increase in the number of D2D users in the hierarchical HetNet. This is due to the frequency reuse, whereas the traditional HetNet shows a constant sum rate. By increasing the number of users in a traditional HetNet, the channel bandwidth per user reduces to accommodate the new users in a fair and uniform manner. However, the sum rate calculated for the increased number of users under fixed-system bandwidth will remain constant as validated by the simulation results. For the hierarchical HetNet, the channel bandwidth for a cellular user decreases, but the D2D communication reuses the channel bandwidth and results in a sum rate enhancement. An interesting cross-over point is observed at 11 milli users/m^2, where single MBS with D2D links shows higher capacity than huge deployments of SBSs with non-D2D links. This cross-over point can be reached at low user density if the number of D2D links is increased further. However, D2D pairs can be exploited opportunistically depending on different factors, for example, shortest distance, channel conditions and common content information.

The overall system gain of the hierarchical HetNet depicted in Figure 9.5 shows significant capacity enhancements compared with the constant sum rate of the traditional HetNet. These capacity gains can further be enhanced by using non-orthogonal spectrum sharing and smart interference management techniques. In such a scenario, the optimum number of D2D pairs can be found, for example, by achieving the target SIR at the desired node (cellular or D2D).

In Figure 9.6, the interference geometry is drawn for a traditional and hierarchical HetNet. Two user densities are simulated: 1 milli user/m^2 and 10 milli users/m^2 (closer to the cross-over point). In both cases, the CDF plot shows significant improvements in terms of required SIR and outage probability. For example, to ensure an outage probability of 10% in case of 1 milli user/m^2, the HetNet with D2D links requires $\gamma_1 = 26.66$ dB less SIR than a traditional HetNet. Similarly, in case of 10 milli users/m^2, the SIR gain $\gamma_{10} = 32.74$ dB was observed.

In the next section, the mathematical analysis to compute the average transmission power of a user in the network is presented.

9.4 Average User Transmission Power Analysis

This section assumes that the mobile users are distributed according to an independent PPP $\Phi = \{L_i, \mu, D_i, P_i^t\}$, where L_i, μ, D_i and P_i^t represent the spatial locations of the users, user intensity per m^2 throughout the network, communication link length and transmission power. For simplicity, the subscript 'i' ($i = s$ or d) refers to small-cell users and D2D communication users, respectively. A distance-based D2D communication mode selection

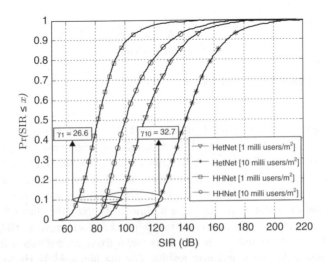

Figure 9.6 Interference Geometry for two user densities

model is considered, where the D2D mode is selected only if (iff) $D < x$, where x denotes the D2D communication link threshold; otherwise, the mobile user communicates through its closest SBS.

Assuming there is only one small-cell per cell coverage area (πR_s^2) and the average number of small-cells per square meter is denoted by μ_s, the radius of a small-cell is given by $R_s = \sqrt{1/\pi\mu_s}$. Hence,

$$\mathcal{P}(D_s \le r) = \frac{r^2}{R_s^2}, \tag{9.18}$$

where $0 \le r \le R_s$. Variable r denotes the distance of a small-cell user from its serving SBS. The PDF of a typical small-cell link length is found by taking the derivative of (9.18) and substituting $R_s = \sqrt{1/\pi\mu_s}$:

$$f_{D_s}(r) = 2\pi\mu_s r. \tag{9.19}$$

Hence, the average transmission power of a small-cell user can be expressed as

$$\mathbb{E}[P_s^t] = P_0\mathbb{E}[D_s^\alpha] = P_0 \int_0^{\sqrt{\frac{1}{\pi\mu_s}}} 2\pi\mu_s r^{\alpha+1} dr = \frac{P_0}{(1+\frac{\alpha}{2})\pi^{\frac{\alpha}{2}}\mu_s^{\frac{\alpha}{2}}}. \tag{9.20}$$

The D2D communication link is assumed to be Rayleigh distributed due to the effect of the user distribution in the network on the D2D communication link length, that is, the larger μ is, the shorter the average D2D communication link distance is.

Recall that D2D communication only takes place if $D_d < x$, where x is the D2D communication link threshold. Thus, the probability of $D_d < x$ is expressed as

$$\mathcal{P}(D_d < x) = 1 - e^{-\pi\mu x^2}. \tag{9.21}$$

Therefore, the PDF of the length of a typical D2D communication link can be expressed as

$$f_{D_d|D_d<x}(r) = \frac{f_{D_d}}{\mathcal{P}(D_d < x)} = \frac{2\pi r\mu e^{-\pi\mu r^2}}{1 - e^{-\pi\mu x^2}}, \tag{9.22}$$

where $0 \le r \le x$. Hence, the average transmission power of a typical D2D communication link is given by

$$\mathbb{E}[P_d^t|D_d < x] = \frac{P_0\mathbb{E}[D_d^\alpha]}{\mathcal{P}(D_d < x)} = \frac{P_0 \int_0^x 2\pi\mu r^{\alpha+1} e^{-\pi\mu r^2} dr}{1 - e^{-\pi\mu x^2}}. \tag{9.23}$$

After some simplifications, (9.23) is expressed as

$$\mathbb{E}[P_d^t|D_d < x] = \frac{P_0 z x^\alpha \, {}_1F_1(a; a+1; -z)}{a(1 - e^{-z})}, \tag{9.24}$$

where $a = \frac{\alpha}{2} + 1$, $z = \pi\mu x^2$ and ${}_1F_1$ denotes the confluent hypergeometric function. The derivation of (9.23) is carried out in Appendix A.

Given that D2D communication only takes place if the intended D2D communication receiver is within the D2D communication range, that is, $D_d < x$, otherwise the closest SBS is used, the average transmission power of a user in the network is given by

$$\mathbb{E}[P_i^t] = \mathcal{P}(D_d < x)\mathbb{E}[P_d^t|D_d < x] + \mathcal{P}(D_d \ge x)\mathbb{E}[P_s^t], \tag{9.25}$$

where

$$\mathcal{P}(D_d \ge x) = 1 - \mathcal{P}(D_d < x) = e^{-\pi\mu x^2}. \tag{9.26}$$

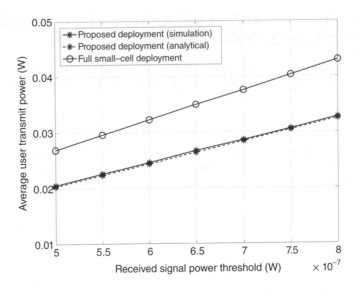

Figure 9.7 Average user transmission power comparison of our proposed deployment against full small-cell deployment

9.4.1 Discussion on Transmission Power Analysis of D2D Users

This section presents the transmission power performance of our proposed network deployment. We only consider the effect of path loss in our simulation. The simulation parameters are summarized in Table 9.1.

A comparison of the simulation and analytical results of the proposed deployment in terms of average user transmission power is illustrated in Figure 9.7. The simulation results are further compared with the results corresponding to a full small-cell deployment network and maximum transmission power. Figure 9.7 shows that the simulation and analytical results match and that the proposed deployment has a considerably lower average user transmission power than the full small-cell deployment. This is attributed to the lower transmission powers of the D2D communications required by the shorter communication link. However, the average user transmission power of both deployments increases as the received signal threshold increases. This is due to the fact that the users must transmit with higher power to overcome the effect of path loss and achieve the minimum received signal power at the receiver (of the D2D communication or SBS). It turns out from the Figure 9.7 that our proposed deployment achieves up to 25% reduction in average user transmission power compared with the full small-cell deployment.

Figure 9.8 shows the average user transmission power saving of the proposed scheme and small-cell deployment. The transmission power saving depicts how much power a typical user is able to conserve by incorporating D2D communication in the network. It can be observed that the average user transmission power saving decreases as the received signal power threshold increases. This is a result of the increased transmission power of the users as the minimum received signal power increases. It turns out that the proposed deployment achieves a higher average user transmission power saving than the full small-cell deployment.

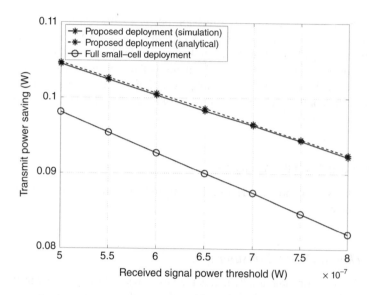

Figure 9.8 Transmission power saving of our proposed deployment against full small-cell deployment

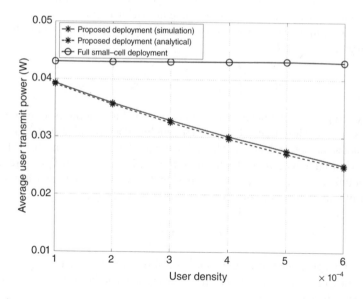

Figure 9.9 Average user transmission power comparison of our proposed deployment against full small-cell deployment versus user density

Figure 9.9 shows the average user transmission power of the proposed scheme against the full small-cell deployment as the user density increases at $P_0 = 0.8$ μW. It can be inferred from the figure that the average transmission power of a small-cell user is constant as the user density increases. This is because the user density does not affect the distance between the user, the SBS and the average transmission power (9.20). On the contrary, as the user density increases, the number of potential D2D communications increases, which results in the mobile users being closer. The shorter link distances lead to the reduction of average user transmission power in the considered deployment as the user density increases as shown in (9.24). Even though the average transmission power of the small-cell user is the same as that of the full small-cell deployment, the incorporation of D2D communications lowers the average transmission power of users.

9.5   Backhaul Energy Analysis

This section analyses the three-tier HetNet in terms of backhaul power consumption and backhaul energy efficiency.

9.5.1   Backhaul Power Consumption

The backhaul power consumption, which is the power needed to carry user traffic to the core network, depends on the type of deployment and the small-cell technology used. D2D communication has no backhaul power requirement, because D2D communication user traffic is not routed to the core network, as the mobile users engage in direct communication without the need for any intermediary node. Therefore, the total backhaul power requirement of the network with D2D communication is simply the backhaul power requirement of the macro-cell BS, and is expressed as [257]

$$P_{\text{BH}}^{\text{macro}} = \left\lceil \frac{1}{\max_{N_{\text{DL}}}} \right\rceil P_s + P_{\text{DL}} + I_{\text{UL}}P_{\text{UL}}, \tag{9.27}$$

where $\max_{N_{\text{DL}}}$ represents the maximum number of downlink interfaces at the macro-cell BS aggregation switch and it is used to compute the number of aggregation switches needed. Variable P_{DL} denotes the power consumed by a downlink interface at the macro-cell aggregation switch, and it is used to receive the backhaul traffic. Variables I_{UL} and P_{UL} represent the total number of uplink interfaces and the power consumption of one uplink interface, respectively. The number of uplink interfaces is given by Skubic and Ericsson [257]

$$I_{ul} = \left\lceil \frac{C_{\text{agg}}}{T_{\text{max}}} \right\rceil, \tag{9.28}$$

where C_{agg} is the aggregate traffic at the macro-cell BS switch(es) and T_{max} is the maximum transmission rate of an uplink switch. The term P_s denotes the power consumption of the aggregation switch, and is expressed as

$$P_s = \Phi P_{\text{max}} + (1 - \Phi) \frac{C_{\text{agg}}}{C_{\text{switch}}^{\text{max}}} P_{\text{max}}, \tag{9.29}$$

where P_{max} is the maximum power consumption of the switch, $C_{\text{switch}}^{\text{max}}$ represents the maximum traffic that the switch can carry, and Φ denotes the weighting factor [257].

We assume that the traffic from the small-cells (femto cells) is routed straight to the core network via the Internet, without going through the aggregation node at the macro-cell BS. The access network of the small-cells is assumed to be a passive optical network (PON). A single fibre cable from the core network, which serves a group of small-cells, is fed into an optical line terminal (OLT), which may be located at the local exchange. A passive curb at the local exchange splits the fibre cable from the OLT into several fibres, each connected to an optical network unit (ONU). Each ONU then serves a single small-cell. The OLTs are connected to the edge routers, which serve as the small-cell gateways for transmission to the core network. The power consumption of the small-cell backhaul is expressed as

$$P_{\mathrm{BH}}^{\mathrm{sc}} = \lceil \frac{N}{K} \left[\frac{P_{\mathrm{router}}}{40} + P_{OLT} \right] + N \cdot P_{\mathrm{ONU}} \tag{9.30}$$

where $K = 4\,\mathrm{Gbps}/C_s$ denotes the number of ONUs that connect to one OLT, C_s represents the total traffic of the small-cells, P_{OLT} denotes the power consumption of the OLT, P_{ONU} denotes the power consumption of the ONU [258] and P_{router} represents the power consumption of the edge router, which can support up to 40 OLTs [259].

9.5.2   Backhaul Energy Efficiency

The backhaul energy efficiency (BEE), which shows the energy utilization of the backhaul technology, is a key performance indicator for future mobile communication systems. BEE expressed as the maximum amount of bits that can be transmitted per joule of energy consumed by the backhaul network, and it is measured in bit/Joule [260]. BEE is important particularly when choosing the type of backhaul technology to use during network planning to bring down the operational expenditure (OPEX) of the network. BEE can be also expressed as

$$\mathrm{BEE} = \frac{C}{P^{\mathrm{net}}}, \tag{9.31}$$

where C is the achievable throughput of the network and P^{net} represents the resultant backhaul power consumption of the network expressed as the sum of the power consumption of the backhaul network and the downlink power consumption:

$$P^{\mathrm{net}} = P_{\mathrm{BH}}^{\mathrm{macro}} + P^{\mathrm{macro}} + P^{\mathrm{D2D}}. \tag{9.32}$$

Variable P^{D2D} represents the total transmission power of the D2D communication users. The total power consumption of the full small-cell network is expressed as

$$P^{\mathrm{net}} = P_{BH}^{\mathrm{macro}} + P_{\mathrm{BH}}^{\mathrm{sc}} + P^{\mathrm{macro}} + N P^{\mathrm{SBS}}, \tag{9.33}$$

where

$$P^{\mathrm{macro}} = \Delta_m P_{\mathrm{MBS}} + P_{\mathrm{MBS},0}, \tag{9.34}$$

and

$$P^{\mathrm{SBS}} = \Delta_s P_{\mathrm{SBS}} + P_{\mathrm{SBS},0}. \tag{9.35}$$

The parameters P^{macro} and P^{SBS} denote the power consumptions of the macro-cell BS and each small-cell BS, respectively. The parameters Δ_m and Δ_s represent the slope of the load-dependent power consumption of the macro-cell BS and small-cell BS, respectively. Variables P_{MBS} and P_{SBS} denote the transmission power of the macro-cell BS and small-cell BSs, respectively. Furthermore, $P_{\mathrm{MBS},0}$ and $P_{\mathrm{SBS},0}$ denote the overhead power consumption of the macro-cell and small-cell BS, respectively [261].

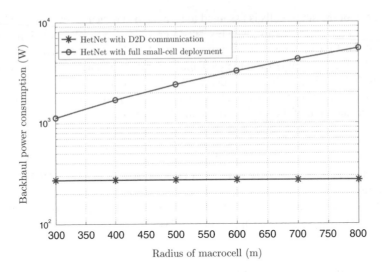

Figure 9.10 Backhaul power consumption comparison of the network with D2D communication against full small-cell deployment

9.5.3 Considerations on Backhaul Energy Efficiency of Hierarchical HetNet

This section compares the performances of the proposed network with D2D communication against the network with full small-cell deployment in terms of the backhaul power consumption and BEE. The simulation parameters are summarized in Table 9.1.

Figure 9.10 depicts the backhaul power consumption of the proposed network with D2D communication against a network with full small-cell deployment by assuming the throughput constant and a varying macro-cell radius from $R_m = 300$ to 800 m. Figure 9.10 indicates that the network with D2D communication presents a significantly lower backhaul power consumption than the network with full small-cell deployment. This is because D2D communication users have no need for any backhaul network to convey their traffic to the core network and only the macro-cell users have their traffic carried by the backhaul network from the macro-cell BS to the core network. On the contrary, the backhaul power requirement of the network with full small-cell deployment increases as the radius of the macro-cell increases. This is due to the increase in the population of small-cells in the network as the macro-cell radius increases and each small-cell has its own backhaul power requirement. It turns out that the network with full small-cell deployment presents about 4–20 times higher backhaul power consumption than that of the network with D2D communication, depending on the radius of the macro cell.

Figure 9.11 illustrates the BEE comparison of the network with D2D communication and the full small-cell deployment. The network radius and the throughput of the macro cell were fixed at $R_m = 500$ m and $C_{agg} = 5$ Mbps, while the total throughput of the network was varied from 10 to 100 Mbps. It can be seen that the BEE of both networks increases as the throughput of the network increases. This is because BEE is a function of the throughput and the total power consumption of the backhaul network. The BEE of the network with D2D communication is at least 260% higher than that of the network with full small-cell deployment. The higher

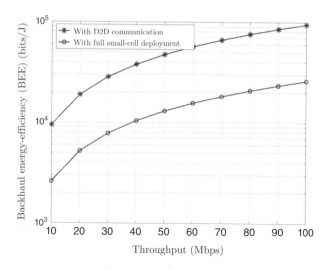

Figure 9.11 Backhaul energy-efficiency comparison of D2D communication against full small-cell deployment for a fixed macro-cell radius $R_m = 500\,\mathrm{m}$

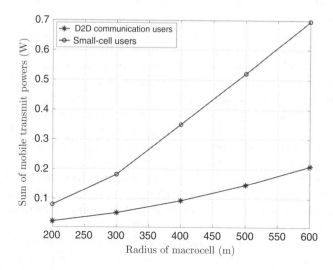

Figure 9.12 Tier 2 uplink sum transmission power comparison of D2D communication against full small-cell deployment

BEE of the D2D communication is due to the lack of backhaul power consumption for the D2D communication users and only the macro-cell users' traffic is backhauled to the core network. However, the backhaul power consumption of each small-cell in the network with full small-cell deployment has to be considered in calculating the BEE, which results in a lower BEE of the network.

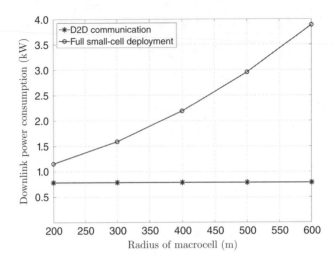

Figure 9.13   Downlink power consumption comparison of D2D communication against full small-cell deployment

Figure 9.12 depicts the total transmission powers of the tier 2 D2D communication users and small-cell users against the macro-cell radius. It turns out that the D2D communication users exhibit a lower transmission power compared to the small-cell users. The lower transmission power is due to the shorter transmitter–receiver link in the D2D communication relative to the small-cell access distance, and the mobile users transmit with just enough power to overcome the effect of path loss via power control. The D2D communication users achieve up to 250% transmission power reduction at a macro-cell radius of 600 m compared to the small-cell users. However, the sum transmission powers of both schemes assume larger values because of the increase in the number of users as a result of the increase in the macro-cell radius.

Figure 9.13 shows the downlink power consumption comparison of the network with D2D communication with the network with the full small-cell deployment at different macro-cell radii. It turns out that the full small-cell network presents a much higher downlink power consumption that increases as the macro-cell radius increases. This is due to the increase in the population of small-cells in the network as the radius of the macro cell increases. Although the downlink power consumption of the network with D2D communication appears to be constant, there is a marginal increase in the downlink power consumption due to the increased user population as the macro-cell radius increases. The network with D2D communication achieves a downlink power consumption reduction of up to 400% at a macro-cell radius of 600 m. Even though the transmission power of the D2D communication users is very low, transmissions over long periods of time (as is the case with mobile multiplayer gaming) may have significant impact on the battery life of the D2D communication terminals.

9.6 Summary

In this chapter, we introduced a three-tier network as a hierarchical HetNet, in which D2D links are established in macro or small-cells. Two scenarios are simulated. The first scenario comprised a HetNet without D2D links, and the second scenario considered a hierarchical HetNet with overlay D2D communication. We used distance-based criteria for mode selection such that the D2D communication mode is selected if the mobile receiver is within the D2D communication range; otherwise, the mobile user connects to the closest SBS. The D2D user density is varied from low to high values to simulate an ultra-dense urban environment. The capacity enhancements have been investigated by comparing the traditional HetNet with the hierarchical HetNet. Simulation results show that the proposed deployment outperforms the full small-cell deployment by reducing the backhaul power consumption of the network, which increases the backhaul energy efficiency of the network. Moreover, the smaller transmitter-to-receiver distance in D2D communications reduces the total uplink transmission power of mobile users. We also derived an analytical expression for the average transmission power of a user in the network. Simulation results show that hierarchical HetNet with D2D communications outperforms the full small-cell deployment in terms of average user transmission power.

Appendix A

Integrating (9.23) leads to

$$\mathbb{E}[P_d^t | D_d < x] = \frac{P_0 \left(\Gamma(\frac{\alpha}{2} + 1) - \Gamma(\frac{\alpha}{2} + 1, \pi\mu x^2) \right)}{(\pi\mu)^{\frac{\alpha}{2}} (1 - e^{-\pi\mu x^2})} \tag{9.36}$$

where $\Gamma(.)$ and $\Gamma(.,.)$ denote the gamma and incomplete gamma functions, respectively. Given that the generalized incomplete gamma function can be decomposed as [262]

$$\Gamma[\tilde{x}, y, \tilde{z}] = \Gamma(\tilde{x}, y) - \Gamma(\tilde{x}, \tilde{z}), \tag{9.37}$$

by setting $y = 0$ makes (9.37) resemble the top part of (9.36). Hence, (9.37) can be expressed as

$$\mathbb{E}[P_d^t | D_d < x] = \frac{P_0 \Gamma(\frac{\alpha}{2} + 1, \pi\mu x^2)}{(\pi\mu)^{\frac{\alpha}{2}} (1 - e^{-\pi\mu x^2})}, \tag{9.38}$$

where $\Gamma(s, q) = \int_0^q t^{s-1} e^{-t} dt$. Using the relationship between the confluent hypergeometric function and gamma incomplete function [263]:

$$_1F_1(b; b + 1; -c) = bc^{-b} \Gamma(b, c), \tag{9.39}$$

(9.38) can be expressed, after some manipulations, in terms of the confluent hypergeometric function as

$$\mathbb{E}[P_d^t | D_d < x] = \frac{P_0 z x^\alpha \, _1F_1(a; a + 1; -z)}{a(1 - e^{-z})}, \tag{9.40}$$

where $a = \frac{\alpha}{2} + 1$ and $z = \pi\mu x^2$.

Appendix B - Simulation Parameters

Table 9.1   Backhaul power consumption simulation parameters

Parameter	Value	Parameter	Value
P_{max} (W)	300	P_{dl} (W)	1
P_{router} (kW)	4	P_{OLT} (W)	100
P_{ONU} (W)	4.69	P_{ul} (W)	2
T_{max} (Gbps)	10	max_{dl}	24
$C_{\text{switch}}^{\text{max}}$ (Gbps)	24	Φ	0.9
P_{MBS} (W)	20	P_{SBS} (W)	0.05
$P_{\text{MBS},0}$ (W)	354.44	$P_{\text{SBS},0}$ (W)	4.8
P_u (W)	0.8	P_0 (μW)	0.8
R_s (m)	25	R_0 (m)	10
Δ_m	21.4	Δ_s	7.5
$\alpha_m = \beta_m$	2.1	$\alpha_s = \beta_s$	1.8
$H_{\text{BS}}(\text{macro})(\text{m})$	25	$H_{\text{BS}}(\text{SBS})(\text{m})$	5
H_u (m)	2	Γ	1
μ (user/m^2)	0.003	λ_c (m)	0.125
Small-cell density (μ_s)	$1/\pi R_s^2$ m^{-2}	User density (μ)	0.0003 m^{-2}
Max. user tx. power (P_i^{max})	0.125 W	Received signal power threshold (P_0)	0.8 μW
Small-cell radius (R_s)	30 m	Coverage area radius (R)	300 m
D2D comm. threshold (x)	20 m	System bandwidth	20 MHz

10

Emerging Device-Centric Communications

Smartphones are equipped with multiple radio interfaces that enable them to access different types of wireless networks, including WLANs, Bluetooth and Zigbee, besides cellular networks. Emerging device-centric systems (DCS) such as devices-to-device communications are considered standard components of future mobile networks, where operators/consumers involve their devices in direct communications to improve the cellular system throughput, latency, fairness and energy efficiency. However, the battery life of the mobile devices involved in such communications is crucial for 5G smartphone users to explore the potential of emerging applications in DCS. It is anticipated that the owners of 5G-enabled smartphones will use their devices to talk, text, e-mail and surf the Internet more often than the customers with 4G smartphones and traditional handsets, which puts a significantly higher demand on the battery life. This chapter introduces a new scheme to support emerging features in DCS, where a device-to-device (D2D)-enabled mobile device (sink device or a content requester) aggregates the radio resources of multiple mobile devices (source devices or content providers) to improve the file transfer latency (FTL), energy efficiency and battery life. This scheme is referred to as devices-to-device (Ds2D) communications. In such a networking setting, this chapter discusses a network-controlled algorithm for optimal selection of source devices and their respective radio interfaces to support green Ds2D communications. Ds2D communications ensure an optimal packet split among the source mobile devices to reduce the FTL and hence to prolong the mobile battery life. Simulation results demonstrate that the proposed optimal packet split scheme guarantees an improvement in the mobile battery life over a wide range of data rate levels in comparison with the random packet split strategy and the traditional D2D communication paradigm between the sink and source mobile devices.

10.1 Introduction

The recent widespread use of mobile Internet complemented by the advent of many smart applications has led to an explosive growth in mobile data traffic over the last few years. This

Green Heterogeneous Wireless Networks, First Edition. Muhammad Ismail, Muhammad Zeeshan Shakir,
Khalid A. Qaraqe and Erchin Serpedin.
© 2016 John Wiley & Sons, Ltd. Published 2016 by John Wiley & Sons, Ltd.

remarkable growing momentum of the mobile traffic will most likely continue on a similar trajectory, mainly due to the emerging need for connecting people, machines and applications in an ubiquitous manner through the mobile devices. Every new release of an iPhone and Android smartphone spurs new applications and services, with advanced display screens to deliver an exceptional quality of experience to the end user. As a result, the current and projected dramatic growth of mobile data traffic necessitates the development of fifth-generation (5G) mobile communications technology. The 5G communications will provide us with the promise of a mobile broadband experience far beyond the current 4G systems. The 5G technology has a broad vision and envisages design targets that include 10–$100 \times$ peak date rate, $1,000 \times$ network capacity, $10 \times$ energy efficiency and 10–$30 \times$ lower latency [264]. In order to achieve these expectations, operators and carriers are planning to leverage emerging device-centric systems (DCS) such as device-to-device (D2D) communications, small-cells and nano and elastic cells to improve the user experience and consequently improve the overall network performance. However, the evolution of mobile devices to support the emerging features in DCS comes at a cost that places stringent demands on the mobile device battery life and energy consumption [265]. Hence, there are considerable market interests on the development and deployment of innovative green and smart solutions to support emerging features in DCS in ultra-dense heterogeneous networks.

10.2   Emerging Device-Centric Paradigms

From 2G to 4G, systems are based on network-centric approaches, but 5G systems will drop this assumption and move toward DCS. It is envisioned that the 5G networks will be mostly deployed for data-centric applications rather than voice-centric applications. The main drivers of DCS are the Internet of things (IoT), machine-to-machine communications and BigData applications, which will exploit the intelligence at the mobile device side to support the emerging device-centric communication paradigms and ensure ubiquitous connectivity.

D2D communication is considered a promising technology to complement the 5G DCS. As shown in Figure 10.1, traditional D2D communications take place among two devices, that is, a pair of devices D_4 and D_5 such that a direct communication link is established between the two mobile devices without any interaction from the BSs or the core of the cellular network. In [266], the authors have provided a literature review on D2D communications, including new insights concerning existing works and emerging protocols. This study includes a review on the inband (underlay or overlay in cellular spectrum) and outband (unlicensed spectrum) integration of D2D communications. In the literature, outband D2D communication uses a cellular interface to set up the connection and the WiFi interface for data transmission between the two devices involved in the D2D communication. Another form of D2D communication involves a pair of devices communicating over multiple interfaces, that is, a pair of devices D_6 and D_7 performing data transmission over both cellular and WiFi interfaces in a D2D set-up (multi-homing D2D pair). Researchers are still formulating the design objectives as optimization problems, but leaving them unsolved due to their NP-hardness. Consequently, most of the proposed algorithms such as the heuristic algorithm [267] and linear/nonlinear/dynamic algorithms [267–269] are subjects open for investigation for new optimal solutions for pairing the devices involved in such communications. D2D communication is also considered as a traffic offloading technology and has received much attention from the operators. However, the

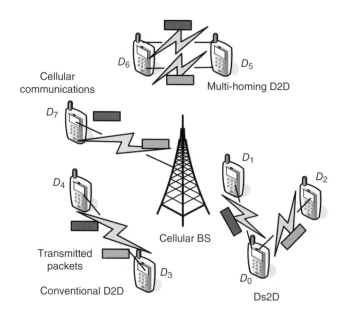

Figure 10.1 Illustration of conventional D2D, multi-homing D2D and Ds2D communication approaches

feasibility of its large-scale implementation and integration into an ultra-dense heterogeneous communication infrastructure is still an open research problem.

10.2.1 Device-to-Device Communication Management

Direct D2D communication between cellular equipments is proposed to increase data rate and extend conventional cellular coverage. In an underlay scheme, the D2D communication may generate interference to the neighbouring cells due to the reuse of the same resources. Therefore, in the underlay approach, D2D links may only exist if they do not harm the SINR at the BSs (uplink) or at the other devices (downlink) in the conventional communication approach. Researchers have proposed different interference management algorithms to increase network capacity [270, 271]. For instance, the authors of [270] proposed that the D2D users monitor the received power of the downlink control signals to control their uplink transmit power below a threshold to avoid high interference to cellular users. If the required transmission power for a D2D link is higher than an interference threshold, then the D2D link is forbidden. One of the proposed solutions for future applications and services in DCS is to reduce the time-average interference power over different networks for both conventional users (communicating through BSs and access points (APs)) and D2D users.

10.2.2 Device-to-Device Communication Architecture

In [272], the authors proposed a new LTE-A-based D2D communication network architecture. They have introduced a new reference point between D2D-enabled devices named 'Di interface' using enhanced radio protocols. The following D2D-specific functionalities are supported by many functions of this interface: (i) the D2D scheme should have the ability to measure the distance between two mobile devices to assess the feasibility of direct connection; (ii) the devices in the D2D architecture should be covered by the eNodeBs to maintain control and signalling and (iii) D2D data transmission between the devices should utilize a physical channel similar to the LTE-A uplink/downlink shared channel.

10.2.3 Device-to-Device Communication Challenges

Some challenges to implement and integrate D2D communications into 5G networks are listed below [266]:

- *Interference Management*; For the reuse of uplink and downlink resources in D2D communications in a small-cell, the D2D mechanism should be designed in a way not to disrupt the cellular network services.
- *Power/Resource Allocation*; The transmission power should be properly regulated so that the D2D transmitter does not interfere with the cellular mobile user communication while maintaining a minimum SINR requirement for the D2D receiver.
- *Channel Measurement/Modulation Format*; D2D communication requires information about the channel gains between D2D pairs, between D2D transmitter and cellular devices and between cellular transmitter and D2D receiver. As the devices are supposed to communicate with both BSs and other peers, it will be very convenient to maintain a common physical layer signalling waveform such as the OFDM modulation.
- *Energy Consumption*; While the energy consumption is a very important issue in D2D communication, it becomes very crucial to propose advanced device discovery, device pairing and D2D communication protocols, which save the battery life of the mobile devices while keeping the required QoS and connectivity.

10.3 Devices-to-Device Communications

The opportunity of enabling multiple radio interfaces including WLANs, Bluetooth and Zigbee, besides cellular networks, is not fully exploited in D2D communications, since the D2D communications take place over a single link between two mobile devices involved in a direct communication. Enabling D2D data transmission between multiple source mobile devices and a sink mobile device over multiple radio interfaces is referred to as devices-to-device (Ds2D) communication. As an example, Figure 10.1 shows that the source mobile devices D_1, D_2 and D_3 are involved in Ds2D communication with a sink mobile D_0. Ds2D communication can take advantage of the diverse resources available at different radio interfaces (e.g. the supporting bandwidth). Aggregating such radio resources at the sink device allows for an improved system performance in terms of the achieved throughput, latency and energy efficiency.

10.3.1 System Model

Consider a system model with a single-sink mobile device and a set of candidate source mobile devices. The sink mobile device is required to download a file (content), which is cached in the source mobile devices. Let $\mathcal{D}$ denote a set of mobile devices that are in the coverage area of a single cellular network base station (BS). Four communication modes can be distinguished in such a network setting, as shown in Fig. 10.1:

- Cellular communications, in which the sink device receives its required file from the cellular BS, as shown in Figure 10.1 for D_7.
- Conventional D2D communications, in which the sink device receives its required file from a single source device $D_s \in \mathcal{D}$ over a single radio interface $n \in \mathcal{N}$, as shown in Figure 10.1 between D_3 and D_4.
- Multi-homing D2D communications, in which the sink device receives its required file from a single source device over multiple radio interfaces, as shown in Figure 10.1 between the source device D_5 and the sink device D_6. For the sake of illustration, assume that D_6 requests a file that consists of 2 packets from D_5. Two data communication links are established between D_5 and D_6, which can take place over the LTE-direct and WiFi-direct radio interfaces of the two devices (besides a third *cellular* link that is established for coordination). On the basis of the achieved data rate over each link (radio interface), different number of packets can be transmitted from D_5 to D_6 on each link. For instance, one data packet is transmitted over the first link and another data packet is transmitted over the second link, as shown in Figure 10.1, assuming equal achieved data rates on each link. Eventually, the sink device D_6 aggregates the received 2 packets to reconstruct the required file.
- Ds2D communications, in which the sink device receives its required (popular) file from multiple source devices over multiple radio interfaces, as shown in Figure 10.1 between the source devices D_1 and D_2, and the sink device D_0. Data communication links are established between each source device and the sink device over different radio interfaces. For instance, data communication can take place between D_1 and D_0 over the LTE-direct radio interface and between D_2 and D_0 over the WiFi-direct radio interface (besides a second *cellular* link that is established between each source device and the sink device for coordination). Again, on the basis of the achieved data rate over each link (radio interface), different number of packets can be transmitted from each source device D_1 and D_2 to D_0. In Fig. 10.1, one data packet is transmitted from D_1 and another data packet is transmitted from D_2 and the sink device D_0 aggregates the received 2 packets to reconstruct the required file.

A network-controlled Ds2D communications approach is considered. Hence, in Ds2D communications, the sink mobile device requests a given (popular) file from the BS and indicates that it can operate in a Ds2D communication mode. The BS broadcasts the file request message to the mobile devices within the sink device proximity. On the basis of the mobile devices feedback, the BS defines a set of candidate source devices that (i) are within the proximity of the sink device, (ii) have a copy of the (popular) file required by the sink device and (iii) are willing to contribute in such a Ds2D communication. Then, the BS selects (from the available candidate source devices) the optimal source devices and their respective radio interfaces that deliver the required file to the sink device in the most energy-efficient manner. After optimal selection of source devices and their respective radio interfaces, the BS coordinates which source device transmits which chunk of the required file. The sink device aggregates the data chunks transmitted by different source devices. This approach can support data hungry applications such as file download or video streaming.

As a first step of research, we consider a system model with a single sink device and a set of candidate source devices. Let $\mathcal{D} = \{D_0, D_1, \ldots, D_S\}$ with D_0 representing the sink device and $D_s \in \mathcal{D} \setminus \{D_0\}$ representing the candidate source devices. Each mobile device $D_s \in \mathcal{D}$ has a set of distinct radio interfaces $\mathcal{N} = \{1, 2, \ldots, N\}$. Radio interface $n \in \mathcal{N}$ in all mobile devices $D_s \in \mathcal{D}$ employs the same access technology. For instance, $n = 1$ represents cellular radio interface in all mobile devices, $n = 2$ represents an LTE direct radio interface, $n = 3$ represents a WiFi direct radio interface and so on. Let x_{ns} be a binary variable that indicates if the sink device D_0 communicates with source device $D_s \in \mathcal{D} \setminus \{D_0\}$ over radio interface $n \in \mathcal{N}$ for data transfer.

The transmission bandwidth that can be supported at radio interface $n \in \mathcal{N}$ for $D_s \in \mathcal{D}$ is denoted by W_{ns}. Each source device D_s communicates with the sink device D_0 over radio interface n using transmission power P_{ns}. Let ρ represent the power amplifier efficiency for each source device. The circuit power consumption Q_{ns} for source device $D_s \in \mathcal{D} \setminus \{D_0\}$ and radio interface $n \in \mathcal{N}$ scales with the transmission data rate R_{ns} via [273]:

$$Q_{ns} = \mu_{ns} + \beta_{ns} R_{ns}, \tag{10.1}$$

where μ_{ns} and β_{ns} are two constants, measured in watts and watts per bit per second (bps). The total power consumption for source device $D_s \in \mathcal{D} \setminus \{D_0\}$ to communicate over its radio interface $n \in \mathcal{N}$ is given by

$$P_{ns}^T = \frac{P_{ns}}{\rho} + Q_{ns}. \tag{10.2}$$

Let L_{ns} and α_{ns} represent the distance and path-loss exponent between the sink device and source device $D_s \in \mathcal{D} \setminus \{D_0\}$, respectively. Denote by κ_{ns} the Rayleigh random variable associated with the channel between the sink device and radio interface $n \in \mathcal{N}$ of source device $D_s \in \mathcal{D} \setminus \{D_0\}$. The channel power gain is given by

$$h_{ns} = \kappa_{ns} L_{ns}^{-\alpha_{ns}}. \tag{10.3}$$

The average channel power gain between the sink device and radio interface $n \in \mathcal{N}$ of source device $D_s \in \mathcal{D} \setminus \{D_0\}$ is denoted by Ω_{ns}.

Each radio interface $n \in \mathcal{N}$ of the sink device suffers from interference imposed by other mobile devices communicating over that specific band. Let $\mathcal{I}_n = \{1, 2, \ldots, I_n\}$ denote the set of mobile devices interfering with the sink device file reception over radio interface $n \in \mathcal{N}$. The distance between the sink device and the source of interference $i \in \mathcal{I}_n$ is denoted by L_{ni} and α_{ni} denotes the path-loss exponent. Let P_{ni} denote the transmission power of interferer $i \in \mathcal{I}_n$ over radio interface $n \in \mathcal{N}$. The interference power over radio interface $n \in \mathcal{N}$ of the sink device is approximated by a Gaussian random variable with zero mean and variance $\sum_{i \in \mathcal{I}_n} P_{ni} L_{ni}^{-\alpha_{ni}}$. The one-sided noise power spectral density is represented by N_0.

10.4   Optimal Selection of Source Devices and Radio Interfaces

In this section, the problem of optimal selection of source devices and radio interfaces is formulated and an algorithm is presented to solve it.

10.4.1 Device Selection Criteria

The selection criterion of a given radio interface $n \in \mathcal{N}$ of source device $D_s \in \mathcal{D} \setminus \{0\}$ is the average achieved energy efficiency η_{ns}, which is a ratio between the average achieved data rate and the average power consumption. Using Shannon's formula, the achieved data rate over radio interface $n \in \mathcal{N}$ of source device $D_s \in \mathcal{D} \setminus \{D_0\}$ is given by

$$R_{ns} = W_{ns} \log_2 \left(1 + \frac{P_{ns} h_{ns}}{\sum_{i \in \mathcal{I}_n} P_{ni} L_{ni}^{-\alpha_{ni}} + W_{ns} N_0} \right). \tag{10.4}$$

The average achieved data rate on the link between the sink device D_0 and source device $D_s \in \mathcal{D} \setminus \{D_0\}$ over radio interface $n \in \mathcal{N}$ is given by [224]

$$\mathbb{E}\{R_{ns}\} = \frac{W_{ns}}{\ln(2)} \exp\left(\frac{\sum_{i \in \mathcal{I}_n} P_{ni} L_{ni}^{-\alpha_{ni}} + N_0 W_{ns}}{\Omega_{ns} P_{ns}} \right)$$

$$\cdot E_1 \left(\frac{\sum_{i \in \mathcal{I}_n} P_{ni} L_{ni}^{-\alpha_{ni}} + N_0 W_{ns}}{\Omega_{ns} P_{ns}} \right), \tag{10.5}$$

where $\mathbb{E}\{\cdot\}$ denotes the expectation and $E_1(x) = \int_0^{+\infty} \exp(-x) x^{-1} \mathrm{d}x$ denotes the exponential integral. From Lemma 2.1 in [224], a lower bound of the average achieved data rate is given by

$$\tilde{R}_{ns} = \frac{W_{ns}}{2} \log_2 \left(1 + \frac{2\Omega_{ns} P_{nsm}}{\sum_{i \in \mathcal{I}_n} P_{ni} L_{ni}^{-\alpha_{ni}} + N_0 W_{ns}} \right). \tag{10.6}$$

Hence, the average achieved energy efficiency on the link between sink device D_0 and source device $D_s \in \mathcal{D} \setminus \{D_0\}$ over radio interface $n \in \mathcal{N}$ is given by

$$\eta_{ns} = \frac{\tilde{R}_{ns}}{\frac{P_{ns}}{\rho} + \mu_{ns} + \beta_{ns} \tilde{R}_{ns}}. \tag{10.7}$$

The objective is to select the source devices and their respective radio interfaces that maximize the total energy efficiency, that is,

$$\max_{x_{ns} \in \{0,1\}} \sum_{D_s \in \mathcal{D} \setminus \{D_0\}} \sum_{n \in \mathcal{N}} x_{ns} \eta_{ns}. \tag{10.8}$$

The total number of links used for data transmission is upper bounded by the maximum number of available radio interfaces N, excluding the cellular radio interface that is used for coordination, that is,

$$\sum_{D_s \in \mathcal{D} \setminus \{D_0\}} \sum_{n \in \mathcal{N}} x_{ns} \leq N - 1. \tag{10.9}$$

Furthermore, only one source device is allowed to communicate with a given radio interface $n \in \mathcal{N}$ of the sink device, that is,

$$\sum_{D_s \in \mathcal{D} \setminus \{D_0\}} x_{ns} \leq 1, \quad \forall n \in \mathcal{N}. \tag{10.10}$$

For Ds2D communications, each source device employs only a single radio interface for data transmission; thus, we have

$$\sum_n x_{ns} \leq 1, \quad \forall D_s \in \mathcal{D} \setminus \{D_0\}. \tag{10.11}$$

The summation over n in (10.11) excludes the cellular radio interface, which is used for coordination.

Hence, the optimal selection of source devices and radio interfaces for green Ds2D communications is obtained by solving the optimization problem

$$\max_{x_{ns} \in \{0,1\}} \sum_{D_s \in \mathcal{D} \setminus \{D_0\}} \sum_{n \in \mathcal{N}} \eta_{ns} \tag{10.12}$$

$$\text{s.t.} \quad (10.9)\text{--}(10.11).$$

10.4.2 Ascending Proxy Auction for Device Selection

One way to solve (10.12) for Ds2D communications is based on the ascending proxy auctions [274]. In this context, each source device $D_s \in \mathcal{D} \setminus \{D_0\}$ defines a set F_s that includes pairs of candidate radio interface and the achieved average energy efficiency over that interface, that is, $F_s = \{(2, \eta_{2s}), \ldots, (n, \eta_{ns}), \ldots, (N, \eta_{Ns})\}$, which excludes the cellular radio interface that is used for coordination. Define one element of F_s by f_s, for example, $f_s = (n, \eta_{ns})$ and a selection f is given by $f = \{f_s \forall D_s \in \mathcal{D} \setminus \{D_0\}\}$, that is, $f = \{(n, \eta_{n1}), (\hat{n}, \eta_{\hat{n}2}), \ldots, (\check{n}, \eta_{\check{n}S})\}$. Each source device ranks F_s based on η_{ns}. Let $\succ_s$ denote a strict preference ordering over F_s based on η_{ns}. All candidate source devices report such a preference order over the cellular radio interface to the cellular BS, which will be in charge of selecting the optimal combination of source devices and radio interfaces.

Let set $\mathcal{F} \subset F_1 \times F_2 \times \cdots \times F_S$ denote a feasible selection set of source devices and their respective radio interfaces that satisfies the constraints in (10.9)–(10.11). The BS can form the feasible selection set $\mathcal{F}$ by considering possible combinations of F_s elements for all $D_s \in \mathcal{D} \setminus \{D_0\}$ (f) and eliminating those combinations that do not follow the constraints in (10.9)–(10.11). For a given source device, if $f_s = \phi$, then device D_s is not selected to contribute to the Ds2D communication session (i.e. $x_{ns} = 0 \forall n \in \mathcal{N}$ for that device D_s). Furthermore, $(\phi_1, \phi_2, \ldots, \phi_S)$ means that no source device contributes to the Ds2D communication session and the sink device receives the requested file from the cellular BS via cellular communication. The cellular BS specifies a preference ordering $\succ_0$ over the set of feasible selection profile $\mathcal{F}$ based on the total average energy efficiency (i.e. $\sum_{D_s \in \mathcal{D} \setminus \{D_0\}} \sum_{n \in \mathcal{N}} \eta_{ns}$).

The ascending proxy auction works over iterations (t) until the optimal selection of source devices and their respected radio interfaces is obtained. Define a bid as the proposed F_s element from devices D_s at iteration t, i.e., f_{0s}^t and $f_0^t = \{f_{0s}^t \forall D_s \in \mathcal{D} \setminus \{D_0\}\}$. Define B_s^t as the set of bids (radio interfaces and average energy efficiencies) offered by source device D_s till iteration t, that is, $B_s^t = \{f_{0s}^{t-1}, f_{0s}^{t-2}, \ldots, f_{0s}^0\}$. Let $B^t = \{B_s^t \forall D_s \in \mathcal{D} \setminus \{D_0\}\}$. The set of available new bids by device D_s is denoted by C_s^t, that is, feasible radio interface and corresponding energy efficiency that have not been offered till iteration t. The optimal selection of source devices and their respective radio interfaces for Ds2D green communication is described by Algorithm 10.4.16, which is executed by the cellular BS. From Theorem 1 in

[274], the selection made by Algorithm 10.4.16 is a stable (NTU-core) selection with respect to the reported preferences.

Algorithm 10.4.16 Optimal Selection of Source Devices and Their Radio Interfaces at the Cellular BS

Initialization: $B_s^0 = \{\phi\}$, $f_{0s}^0 = \phi$ and $J = 1$;
while $J = 1$ **do**
  for $D_s \in \mathcal{D} \setminus \{D_0\}$ **do**
    $C_s^t = F_s - \{f_s | f_s \succ_s \phi_s\} - B_s^{t-1}$;
  end for
  Any D_s with $f_{0s}^t = \phi$ and $C_s^t \neq \phi$
  $B_s^t = B_s^{t-1} \cup \{\max C_s^t\}$;
  for All D_s with $f_{0s}^t \neq \phi$ or $C_s^t = \phi$ **do**
    $B_s^t = B_s^{t-1}$;
  end for
  if $B^t = B^{t-1}$ **then**
    $J = 0$;
  else
    $\mathcal{F}^t = \mathcal{F} \cap \{ \bigcap_{D_s \in \mathcal{D} \setminus \{D_0\}} \{f | f_s \in B_s^t\}\}$;
    $f_0^{t+1} = \max \mathcal{F}^t$;
    $t \longleftarrow t + 1$;
  end if
end while
Output: f_0^t.

In Algorithm 10.4.16, each source device first updates its new available bids that can be offered in iteration t. If there exists a source device with $f_{0s}^t = \phi$ and still has new bids to offer (i.e. $C_s^t \neq \phi$), the source device will offer the most preferred radio interface to participate in the Ds2D communication (the preference order here is based on the source device most energy-efficient radio interface). The source device also updates the set of bids offered until iteration t (B_s^t). All other devices make no new bid at this iteration. The BS updates the set of feasible bids at the current iteration t ($\mathcal{F}^t$) and then selects the most energy-efficient set of source devices and radio interfaces (the selection here is made based on the total average energy efficiency $\sum_{D_s \in \mathcal{D} \setminus \{D_0\}} \sum_{n \in \mathcal{N}} \eta_{ns}$).

10.4.3 Discussions on Device and Radio Interface Selection

This section presents comparative simulation results for green Ds2D, multi-homing D2D and conventional D2D communications. The optimal selection of source devices and their respected radio interfaces for the Ds2D is implemented using Algorithm 10.4.16. For conventional D2D communications, only the source device and radio interface offering the maximum energy efficiency η_{ns} are selected for data transfer. For multi-homing D2D, the source device achieving maximum total (sum) energy efficiency across all its radio interfaces is selected for data transfer. All mobile devices have two radio interfaces besides the cellular

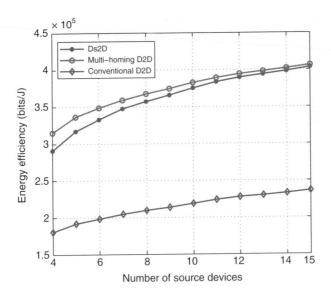

Figure 10.2   Achieved average energy efficiency versus the number of candidate source devices

radio interface (i.e. $N = 3$). In all three modes, coordination is established over the cellular radio interface ($n = 1$) and data transfer can take place over the other radio interfaces ($n = 2$ and 3). The candidate source devices are uniformly distributed within the proximity of [50,100] m away from the sink device. The supporting bandwidth for the radio interfaces used for data transmission are $W_{2s} = 1\,\text{MHz}$ and $W_{3s} = 5\,\text{MHz}$. Each radio interface of the sink device is subject to a random number of interferers uniformly distributed in the range [5,10]. The interferers are assumed to be close to the sink device (for a worst-case scenario), that is, uniformly distributed within the proximity of [50, 60] m away from the sink device. The transmission power is 100 mW for P_{ns} and P_{is}. The power amplifier drain efficiency is 35%. The circuit power constants are $\mu_{2s} = 50\,\text{mW}$, $\mu_{3s} = 75\,\text{mW}$, $\beta_{2s} = 10^{-6}\,\text{W/bps}$ and $\beta_{2s} = 5 \times 10^{-6}\,\text{W/bps}$. The path-loss exponent equals 4 for α_{ns} and α_{ni}, and $N_0 = -174\,\text{dBm/Hz}$.

Fig. 10.2 shows the achieved average energy efficiency versus the number of candidate source devices. With more candidate source devices, a better energy efficiency can be achieved due to the diverse channel conditions among the candidate source devices and the sink device. Both Ds2D and multi-homing D2D communications exhibit an improved energy efficiency performance compared with the conventional D2D communication (up to 70% improvement in energy efficiency). This is mainly due to the aggregated resources at the sink device from multiple radio interfaces, which allows for higher achieved throughput and hence improved energy efficiency. Such an improvement is also due to spatial diversity as some differences are expected in the channel conditions among the sink device and different source devices for Ds2D communications. As shown in Fig. 10.2, Ds2D communications exhibit a closer performance to multi-homing D2D communications as the number of candidate source devices increases. This is due to the higher probability of having more than one source device with good channel conditions with the sink device. While Fig. 10.2 shows a slightly improved performance for multi-homing D2D over Ds2D communications in terms of the total energy

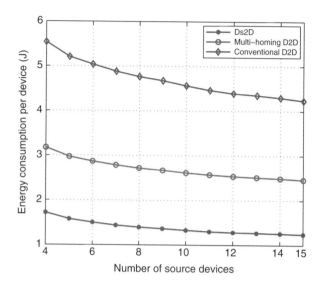

Figure 10.3   Energy consumption per source device to transfer a 1-Mbit file versus the number of candidate source devices

efficiency, the next result shows that Ds2D communications is an attractive alternative as it exhibits a much lower energy consumption per source device. Such an option motivates source devices to contribute to D2D communications.

Fig. 10.3 shows the average energy consumption performance per source device to transfer a 1 Mbit-file to the sink device versus the number of candidate source devices. The worst energy consumption performance per source device is for the conventional D2D communications approach, since only one radio interface is used for data transfer, which results in a longer latency to transfer the file to the sink device, and that results in a higher energy consumption. On the contrary, Ds2D communications exhibit the least energy consumption per source device (up to 70% compared with the conventional D2D communications and up to 50% compared with the multi-homing D2D communications). This is mainly because Ds2D communications split the total energy consumption burden over different source devices contributing to the file transfer, while multi-homing D2D communications relies on a single source device for file transfer, which incurs a higher energy consumption to activate all radio interfaces and transmit across them. With more available radio interfaces at the sink device, additional energy saving is expected per source device when compared with multi-homing D2D, as more source devices will be involved in the file transfer.

10.5   Optimal Packet Split among Devices

After optimal selection of source mobile devices and their respective radio interfaces, the BS coordinates with the source mobile devices to transfer the desired data packets to the sink mobile device in a distributed manner. The sink mobile device aggregates the data packets transmitted by different source mobile devices to reconstruct the required file. This approach can support data hungry applications such as file download or video streaming of a popular content.

The optimal packet split algorithm should specify the packet distribution ratio among the source devices based on the achieved data rates over their respective radio interfaces. Consider that the desired file has P long data packets, which should be transmitted from the source mobile devices (e.g. D_1 and D_2 as shown in Figure 10.1) to the sink mobile device (D_0) over a set of two different radio interfaces $\mathcal{N} = \{1, 2\}$, as shown in Figure 10.1. Let $0 < \alpha_{\text{opt}} \leq 1$ denote the optimal packet split ratio (OPSR) that splits the requested file into two sets of data packets based on the achieved data rate for each selected source device. Set 1 of data packets contains $\alpha_{\text{opt}} P$ data packets that are transmitted by source mobile device D_1 through radio interface $n = 1$. Similarly, set 2 of data packets contains $(1 - \alpha_{\text{opt}}) P$ data packets that are transmitted by the source mobile device D_2 through radio interface $n = 2$. The sink mobile device receives the packets from both source mobile devices simultaneously over two different radio interfaces ($n = 1$ and 2) and combines them to restore the requested file.

The two source mobile devices D_1 and D_2 transmit with different data rates R_1 and R_2, respectively, depending on the SINR of each source mobile device at the corresponding radio interface.[1] The file transfer latency t at the sink mobile device is defined as the duration required to transfer the desired data packets from all source mobile devices to the sink mobile device by aggregating the multiple radio resources, and is given by

$$t = \max_{n \in \mathcal{N}} \left\{ \frac{P_n b}{R_n} \right\} \quad (\text{seconds}), \tag{10.13}$$

where P_n denotes the number of data packets transmitted over the nth radio interface (using α, we have $P_1 = \alpha P$ and $P_2 = (1 - \alpha) P$), R_n denotes the data rate over the nth radio interface and $b = 1,500 \times 8$ denotes the number of bits per data packet. It is assumed that each data packet contains 1,500 bytes. From (10.13), the file transfer latency is minimum if all source devices complete their data transmissions at the same time. Hence, the main rationale behind the search of α_{opt} is to ensure that the source devices involved in Ds2D communications complete the file transfer at the same time such that the sink mobile device does not have to wait for one source mobile device to complete the transmission of its assigned data packets, which elongates the communication session and leads to a higher energy consumption. Thus, α_{opt} can be found by solving $(\alpha_{\text{opt}} P B)/(R_1) = ((1 - \alpha_{\text{opt}}) P B)/(R_2)$.

In order to evaluate the effectiveness of the proposed optimal packet split strategy over the two radio interfaces of two source mobile devices, we consider an average monthly data usage capability for each mobile subscriber of about 2.5 GB with the daily download capability of 80 MB. Given the fact that each data packet has 1,500 bytes, the file (requested content) has $P = 55K$ data packets. The average data rate achieved for the second source mobile device (D_2) over radio interface $n = 2$ is assumed to be 1.646 Mbps. Moreover, the average achieved data rate for the first source device (D_1) over radio interface ($n = 1$) is varied for performance evaluation. Figure 10.4 shows the optimal packet split between a pair of source devices (D_1, D_2) over two radio interfaces $\mathcal{N} = \{1, 2\}$ for different data rates achieved by D_1. The optimal packet split algorithm divides the data packets among the two source mobile devices based on the achieved data rate for each source mobile device. This is mainly because the optimal packet split algorithm ensures the same FTL at each source mobile device, as

---

[1] The relation between the SINR and the achieved data rates is adopted from [275, Table II].

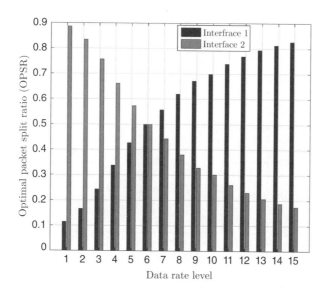

Figure 10.4   Optimal packet split over two interfaces of two source mobile devices vs. range of data rate levels

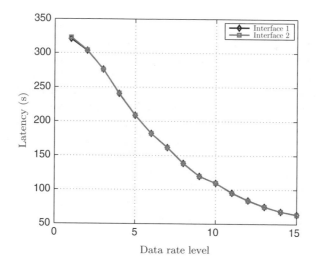

Figure 10.5   Latency of transferring the requested file to the sink mobile device over two radio interfaces of two source mobile devices by exploiting the optimal packet split

shown in Figure 10.5. It turns out that the FTL is dominated by the device suffering from the maximum FTL.

Another performance evaluation criterion is the relative percentage reduction in FTL (i.e. relative gain) as compared with the transmission over the conventional D2D communication paradigm, where only one source mobile device transmits the complete file to the sink mobile device, that is, direct D2D communication between a pair of devices

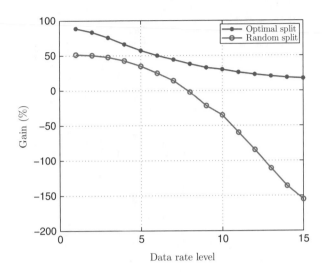

Figure 10.6 Relative gain in file transfer latency (FTL) over Ds2D communication with optimal packet split and random packet split in comparison with direct D2D communication

(D_1 and D_0). The performance of the optimal packet split algorithm is evaluated against a random packet split benchmark. The random packet split benchmark algorithm randomly divides the file among the two source mobile devices and each source mobile device transfers the packets to the sink mobile device that combines both sets to restore the requested file. It can been seen clearly from Figure 10.6 that Ds2D transmission with optimal packet split has a lower transmission FTL than the conventional D2D paradigm (there is always a gain, which ranges from 88% to 30% FTL reduction). Moreover, the optimal packet split is necessary for performance improvement in the Ds2D paradigm, as the random packet split can have an FTL performance worse than the conventional D2D paradigm (gain is below 0% for the data rate levels contained in the range 8–15). Furthermore, as shown in Figure 10.6, with the increase in the data rate level for D_1, the achieved gain is reduced. This is mainly because with high data rates achieved for D_1, a single transmission link (between D_1 and D_0) is already sufficient to achieve a lower FTL than the Ds2D communications (among D_1, D_2 and D_0).

10.6   Green Analysis of Mobile Devices

In this subsection, we present simulation results to show the performance of the proposed Ds2D communication under optimal packet split and random packet split schemes. The energy consumption of Ds2D communication is compared with the D2D communication scenario when the sink mobile device receives the complete file from only one of the source mobile devices over the direct communication link. As an example, consider a source mobile device such that its battery holds a charge of $I_{batt} = 1,440$ mAh with $P_{batt} = 5.45$ Wh [265].

10.6.1   Energy Consumption of Mobile Devices

Energy consumption of the source mobile device for transferring a file to a sink mobile device can be determined as follows:

$$E_{\text{batt}} = \frac{P_{\text{batt}}\, t}{3{,}600} \quad (\text{Wh}), \tag{10.14}$$

where t is the FTL per source mobile device measured in seconds to transfer the desired content to the sink mobile device. Figure 10.7 shows the energy consumption per source device involved in transferring optimally the assigned data packets out of the file of size 80 MB (or equivalently $P = 55\text{K}$ data packets) to a sink device over the range of date rate levels. Compared with the direct D2D communication between a pair of devices, the proposed Ds2D communication offers reduced energy consumption per source mobile device, since each source device only transmits a fraction of packets of the requested file. However, the energy consumption of the source mobile devices involved in Ds2D communication with an optimal packet split scheme outperforms the energy consumption of the source mobile devices that adapt the random packet split scheme. Moreover, at lower date rate levels, the energy consumption of source mobile devices involved in Ds2D with an optimal packet split scheme is significantly reduced in comparison with the energy consumption of the source mobile devices involved in Ds2D communication with a random packet split scheme and traditional D2D communications. The improvement is due to the fact that the source mobile device is engaged with the sink mobile device for a relatively longer duration at a lower data rate level to complete the transfer of the required file under direct D2D communications in comparison with the source mobile devices involved in Ds2D communication. As an example, at a rate level 2, that is, $R_2 = 328.2$ kbs, the source devices can achieve 85% reduction in energy consumption under the optimal packet split scheme and 51% reduction under the random packet split scheme in comparison with the source device involved in D2D communications. As shown in Figure

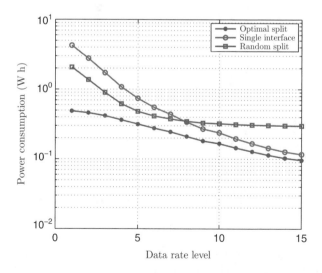

Figure 10.7   Power consumption (Wh) of source devices versus range of achieved data rate

10.7, Ds2D communications with optimally assigned data packets exhibit a closer performance to D2D communications at higher data rate levels. This is due to the fact that with the high data rates exhibited by the source mobile device, a single transmission link (e.g. between D_1 and D_0) is already sufficient to achieve low FTL as compared with the Ds2D communications.

10.6.2   Electricity Cost for Mobile Charging

Reducing energy consumption of mobile devices lowers the electricity cost for charging devices, and thereby, results in financial cost savings to the consumers if the energy savings offset any additional costs for implementing an energy-efficient framework. The monthly cost of electricity that is associated with the implementation of Ds2D communications is calculated by assuming 1 kWh = 12 cents and it assumes the expression

$$\text{Cost} = \frac{E_{\text{batt}}}{1,000 \times 100} \times 12 \quad (\text{USD/month}). \tag{10.15}$$

Figure 10.8 shows the monthly electricity cost associated with the energy savings achieved per source mobile device for the transfer of a 80-MB file over the considered device-centric framework. It can be seen clearly that at an average price of 12 cent/kWh, the mobile device costs approximately 150 USD in addition to the monthly electricity bill of the consumers who assume a pair of devices that are involved in D2D communications with an average daily data usage or file/content transfer of size 80 MB. On the contrary, the electricity cost decreased to 70 and 19 USD, when the devices are involved in Ds2D communication with random packet split and optimal packet split schemes, respectively, assuming the same amount of daily data transfer. In the presence of high data rates achieved by the source mobile devices, a single

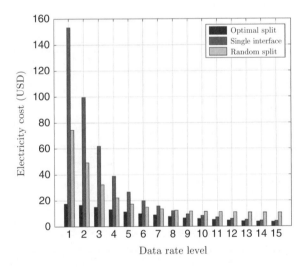

Figure 10.8   Monthly electricity cost for an average download of a file with a size of 80 MB over Ds2D communications with optimal and random packet split schemes and traditional D2D communication for the range of achieved data rate levels

transmission link (e.g. between D_1 and D_0) is sufficient to achieve low FTL as compared with the Ds2D communications. This fact explains the close performance of Ds2D and D2D communications at high data rate levels.

10.6.3   Battery Life of Mobile Devices

Energy efficiency has been recently marked as one of the alarming bottlenecks in the telecommunications growth paradigm mainly due to two major reasons, namely (i) slowly progressing battery technology [276] and (ii) dramatically varying global climate [9]. A recent survey reports that up to 60% of the mobile users in China complained that the battery consumption is the greatest hurdle while using 4G services [277]. Emerging device-centric frameworks can offer a longer battery life, while consumers can enjoy high data rate 5G services and applications. The battery life or battery capacity can be calculated from the input current rating of the battery and the load current of the battery charging circuit [265]. Battery life will be high when the load current is low and vice versa. The capacity of battery is given by DigiKey Electronics [278]

$$\text{Battery Life} = \frac{I_{\text{batt}}}{I_{\text{load}}} \times 0.70 \quad \text{(hours)}, \qquad (10.16)$$

where I_{batt} is the battery capacity in mAh, and I_{load} is the load current drawn by the source mobile device for transferring the file to the sink mobile device. Here, the factor 0.70 represents external factors that can affect the mobile device battery life [278]. Figure 10.9 shows the mobile battery life (h) over the range of data rate levels for a mobile device involved in D2D and Ds2D communications. Overall, as the data rate level increases, the FTL is decreased, and hence the battery life is prolonged. However, it can be seen clearly that the battery life of a mobile device involved in Ds2D communications with an optimal packet split scheme is

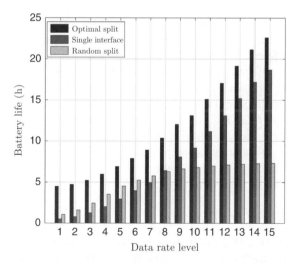

Figure 10.9   Average improvement in battery life of source devices over Ds2D communications with optimal and random packet split and traditional D2D communications for a range of data rate levels

significantly higher than the battery life of a mobile device involved in Ds2D communication via a random packet split scheme and traditional D2D communication. Moreover, the battery life of the mobile device involved in Ds2D communication and that assumes the random packet split scheme degrades at high data rate levels, since the FTL performance of the random packet split scheme is worse than that of the traditional D2D communications (as can be seen from Figure 10.5).

10.7   Some Challenges and Future Directions

In general, Ds2D communications can be established among any sink mobile device and multiple source mobile devices over N multiple radio interfaces. Selection of source mobile devices is highly dependent on the availability of the file or content and its close proximity with the sink mobile device. As discussed earlier, coordination among the involved mobile devices is required for the successful implementation of Ds2D communication and optimal distribution of the desired content (data packets) among the source mobile devices. There are two possible implementation approaches to achieve the coordination among the mobile devices and set up Ds2D communications, as described below.

10.7.1   Centralized Ds2D Set-up

Under the centralized Ds2D set-up, cloud radio access networks (CRAN) perform source mobile device selection, Ds2D link establishment and data packet distribution among the source devices with a limited or full supervision of cellular network. Devices involved in Ds2D communication perform full or limited information exchange and signalling with the cellular network using the LTE–Uu interface (i.e. cellular link). Since the cellular interface for all devices is reserved for information exchange and signalling, data transmission can be established between a sink mobile device and source devices over $N - 1$ radio interfaces. Therefore, the devices have at least two active interfaces (cellular interface for control and an additional radio interface for data transmission). Mobility of the devices involved in Ds2D communication, interference management and content availability is considered as an advantage to integrate Ds2D communications under the centralized CRAN-enabled cellular system. Inter-network caching plays an important role to efficiently exploit the benefits of Ds2D communications. However, the centralized approach imposes additional challenges to the fronthaul requirements such as high data rate and latency due to the information exchange and signalling overheads between the devices involved in Ds2D communications. Moreover, devices cannot establish Ds2D communication links without full or limited intervention and approval of the request by the cellular BS.

10.7.2   Decentralized Ds2D Set-up

Under the decentralized Ds2D set-up, devices involved in Ds2D communications can exchange control signalling for selection of source mobile devices, Ds2D communication establishment and content distribution among the devices without any intervention from the cellular BS. Therefore, devices can establish Ds2D communications over a relatively short time period under a decentralized system in comparison with the time required to set up Ds2D links in a centralized manner. The cellular network does not have any supervision over the functionalities used by the devices involved in Ds2D communications, such as resource allocation and interference management. Devices can use the PC5 interface, which is allocated by the LTE standard for device discovery and Ds2D communication between users. Moreover, the fronthaul requirements can be relaxed due to the reduced signalling information exchange between the devices and access network. Long-term availability of the desired content due to sustainability of connection and mobility is one of the challenges to integrate Ds2D communications in a decentralized manner.

10.8   Summary

Smartphones will play an important role to enable device-centric communication paradigms in 5G networks, such as D2D communications. This chapter focused on the implementation perspectives of such a device centric architecture, including energy consumption and battery life aspects of the devices involved in communication. A new device-centric scheme, Ds2D communication was discussed, and it incorporates several source devices and multiple radio interfaces for data transfer to the sink device. An optimal algorithm for source device and radio interface selection was presented based on the ascending proxy auctions mechanism. The proposed mechanism achieves a higher energy efficiency compared with the conventional D2D communications approach and a lower energy consumption per source device compared with the multi-homing D2D communications approach. The proposed Ds2D communication scheme guarantees an optimal data packet distribution among the source mobile devices and it ensures improvements in the file transfer latency, energy consumption and battery life of the source mobile devices involved in communication. Simulation results evaluated the quantitative gains as exhibited by the traditional D2D and Ds2D communications via random data packet distributions. It illustrated that performance metrics associated with the source mobile devices such as file transfer latency, energy consumption and battery life can be effectively optimized through an optimal packet split strategy among the source mobile devices and their respective radio interfaces involved in Ds2D communications in DCS.

References

[1] M. Ismail and W. Zhuang, Cooperative Networking in a Heterogeneous Wireless Medium, *Springer Briefs in Computer Science*, Springer-Verlag, New York, April 2013.

[2] T. Chen, Y. Yang, H. Zhang, H. Kim, and K. Horneman, "Network energy saving technologies for green wireless access networks," *IEEE Wireless Commun.*, vol. 8, no. 5, pp. 30–38, Oct. 2011.

[3] D. Cavalcanti, D. P., Agrawal, C. Cordeiro, B. Xie, and A. Kumar "Issues in integrating cellular networks, WLANs, and MANETs: a futuristic heterogeneous wireless network," *IEEE Wireless Commun. Mag.*, vol. 12, no. 3, pp. 30–41, Jun. 2005.

[4] M. Ismail and W. Zhuang, "A distributed multi-service resource allocation algorithm in heterogeneous wireless access medium," *IEEE J. Sel. Areas Commun.*, vol. 30, no. 2, pp. 425–432, Feb. 2012.

[5] M. Ismail, A. Abdrabou, and W. Zhuang, "Cooperative decentralized resource allocation in heterogeneous wireless access medium," *IEEE Trans. Wireless Commun.*, vol. 12, no. 2, pp. 714–724, Feb. 2013.

[6] M. Ismail and W. Zhuang, "Decentralized radio resource allocation for single-network and multi-homing services in cooperative heterogeneous wireless access medium," *IEEE Trans. Wireless Commun.*, vol. 11, no. 11, pp. 4085–4095, Nov. 2012.

[7] Z. Hasan, H. Boostanimehr, and V. K. Bhargava, "Green cellular networks: a survey, some research issues and challenges," *IEEE Commun. Surv. Tutorials*, vol. 13, no. 4, pp. 524–540, Sept. 2011.

[8] C. Han, T. Harrold, S. Armour, and I. Krikidis, "Green radio: radio techniques to enable energy-efficient wireless networks," *IEEE Commun. Mag.*, vol. 49, no. 6, pp. 46–54, Jun. 2011.

[9] Y. Chen, S. Zhang, S. Xu, and G. Y. Li, "Fundamental trade-offs on green wireless networks," *IEEE Commun. Mag.*, vol. 49, no. 6, pp. 30–37, Jun. 2011.

[10] H. Bogucka and A. Conti, "Degrees of freedom for energy savings in practical adaptive wireless systems," *IEEE Commun. Mag.*, vol. 49, no. 6, pp. 38–45, Jun. 2011.

[11] S. Mclaughlin, P. M. Grant, J. S. Thompson, and H. Haas, "Techniques for improving cellular radio base station energy efficiency," *IEEE Wireless Commun.*, vol. 18, no. 5, pp. 10–17, Oct. 2011.

[12] M. Ismail and W. Zhuang, "Network cooperation for energy saving in green radio communications," *IEEE Wireless Commun.*, vol. 18, no. 5, pp. 76–81, Oct. 2011.

[13] GeSI/The Climate Group, *SMART 2020: Enabling the Low Carbon Economy in the Information Age*, Global e-Sustainability Initiative, Brussels, 2008.

Green Heterogeneous Wireless Networks, First Edition. Muhammad Ismail, Muhammad Zeeshan Shakir, Khalid A. Qaraqe and Erchin Serpedin.
© 2016 John Wiley & Sons, Ltd. Published 2016 by John Wiley & Sons, Ltd.

[14] L. Suarez, L. Nuaymi, and J. M. Bonnin, "An overview and classification of research approaches in green wireless networks," *EURASIP J. Wireless Commun. Networking*, vol. 142, pp. 1–18, April 2012.

[15] K. Pentikousis, "In search of energy-efficient mobile networking," *IEEE Commun. Mag.*, vol. 48, no. 1, pp. 95–103, Jan. 2010.

[16] G. Miao, "Energy-efficient uplink multi-user MIMO," *IEEE Trans. Wireless Commun.*, vol. 12, no. 5, pp. 2302–2313, May 2013.

[17] E. Oh, B. Krishnamachari, X. Liu, and Z. Niu, "Toward dynamic energy-efficient operation of cellular network infrastructure," *IEEE Commun. Mag.*, vol. 49, no. 6, pp. 56–61, Jun. 2011.

[18] Y. S. Soh, T. Q. S. Quek, M. Kountouris, and H. Shin, "Energy efficient heterogeneous cellular networks," *IEEE J. Sel. Areas Commun.*, vol. 31, no. 5, pp. 840–850, May 2013.

[19] I. Humar, X. Ge, L. Xiang, M. Jo, M. Chen, and J. Zhang, "Rethinking energy efficiency models of cellular networks with embodied energy," *IEEE Network*, vol. 25, no. 2, pp. 40–49, April 2011.

[20] D. Feng, C. Jiang, G. Lim, L. J. Cimini, G. Feng, and G. Y. Li, "A survey of energy-efficient wireless communications," *IEEE Commun. Surv. Tutorials*, vol. 15, no. 1, pp. 167–178, Feb. 2013.

[21] S. Tombaz, A. Vastberg, and J. Zander, "Energy- and cost-efficient ultra-high-capacity wireless access," *IEEE Wireless Commun.*, vol. 18, no. 5, pp. 18–24, Oct. 2011.

[22] L. M. Correia, D. Zeller, O. Blume, and D. Ferling, "Challenges and enabling technologies for energy aware mobile radio networks," *IEEE Commun. Mag.*, vol. 48, no. 11, pp. 66–72, Nov. 2010.

[23] X. Ma, M. Sheng, J. Li, and Q. Yang, "Concurrent transmission for energy efficiency of user equipment in 5G wireless communication networks," *Science China. Information Sciences*, vol. 59, no. 2, pp 1–15, Feb. 2016

[24] G. Miao, N. Himayat, Y. Li, and A. Swami, "Cross-layer optimization for energy-efficient wireless communications: a survey," *Wiley J. Wireless Commun. Mobile Comput.*, vol. 9, pp. 529–542, March 2009.

[25] M. Ismail, W. Zhuang, and S. Elhedhli, "Energy and content aware multi-homing video transmission in heterogeneous networks," *IEEE Trans. Wireless Commun.*, vol. 12, no. 7, pp. 3600–3610, July 2013.

[26] S. Bu, F. R. Yu, Y. Cai, and X. P. Liu, "When the smart grid meets energy-efficient communications: green wireless cellular networks powered by the smart grid," *IEEE Trans. Wireless Commun.*, vol. 11, no. 8, pp. 3014–3024, Aug. 2012.

[27] M. Ismail, W. Zhuang, E. Serpedin, and K. Qaraqe, "A survey on green mobile networking: from the perspectives of network operators and mobile users," *IEEE Commun. Surv. Tutorials*, vol. 17, no. 3, pp. 1535–1556, Nov. 2014.

[28] G. Miao, N. Himayat, G. Y. Li, and S. Talwar, "Low-complexity energy-efficient scheduling for uplink OFDMA," *IEEE Trans. Commun.*, vol. 60, no. 1, pp. 112–120, Jan. 2012.

[29] S. G. Colavolpe, D. Saturnino, and A. Zapone, "Potential games for energy-efficient power control and subcarrier allocation in uplink multicell OFDMA systems," *IEEE J. Sel. Areas Signal Process.*, vol. 6, no. 2, pp. 89–103, April 2012.

[30] O. Onireti, F. Heliot, and M. A. Imran, "On the energy efficiency-spectral efficiency trade-off in the uplink of CoMP system," *IEEE Trans. Wireless Commun.*, vol. 11, no. 2, pp. 556–561, Feb. 2012.

[31] O. Galinina, S. Andreev, A. Turlikov, and Y. Koucheryavy, "Optimizing energy efficiency of a multi-radio mobile device in heterogeneous beyond-4G networks," *Perform. Eval.*, vol. 78, pp. 18–41, Jun. 2014.

[32] K. Ying, H. Yu, and H. Luo, "Inter-RAT energy saving for multicast services," *IEEE Commun. Lett.*, vol. 17, no. 5, pp. 900–903, May 2013.

[33] X. Ma, M. Sheng, and Y. Zhang, "Green communications with network cooperation: a concurrent transmission approach," *IEEE Commun. Lett.*, vol. 16, no. 12, pp. 1952–1955, Dec. 2012.

[34] Z. Niu, Y. Wu, J. Gong, and Z. Yang, "Cell zooming for cost-efficient green cellular networks," *IEEE Commun. Mag.*, vol. 48, no. 11, pp. 74–79, Nov. 2010.

[35] K. Son, H. Kim, Y. Yi, and B. Krishnamachari, "Base station operation and user association mechanisms for energy-delay tradeoffs in green cellular networks," *IEEE J. Sel. Areas Commun.*, vol. 29, no. 8, pp. 1525–1536, Sept. 2011.

[36] X. Zhang and P. Wang, "Optimal trade-off between power saving and QoS provisioning for multicell cooperation networks," *IEEE Wireless Commun.*, vol. 20, no. 1, pp. 90–96, Feb. 2013.

[37] T. Han and N. Ansari, "On optimizing green energy utilization for cellular networks with hybrid energy supplies," *IEEE Trans. Wireless Commun.*, vol. 12, no. 8, pp. 3872–3882, May 2013.

[38] Z. Niu, "TANGO: traffic-aware network planning and green operation," *IEEE Wireless Commun.*, vol. 18, no. 5, pp. 25–29, Oct. 2011.

[39] G. Auer, V. Giannini, C. Desset, and I. Godor, "How much energy is needed to run a wireless network?" *IEEE Wireless Commun.*, vol. 18, no. 5, pp. 40–49, Oct. 2011.

[40] A. da Silva, M. Meo, and M. Marsan, "Energy-performance trade-off in dense WLANs: a queuing study," *Comput. Networks*, vol. 56, pp. 2522–2537, March 2012.

[41] L. B. Le, D. Niyato, E. Hossain, D. I. Kim, and D. T. Hoang, "QoS-aware and energy-efficient resource management in OFDMA femtocells," *IEEE Trans. Wireless Commun.*, vol. 12, no. 1, pp. 180–194, Jan. 2013.

[42] F. Liu, K. Zheng, W. Xiang, and H. Zhao, "Design and performance analysis of an energy-efficient uplink carrier aggregation scheme," *IEEE J. Sel. Areas Commun.*, vol. 32, no. 2, pp. 197–207, May 2013.

[43] N. Mastronarde and M. van der Schaar, "Fast reinforcement learning for energy-efficient wireless communication," *IEEE Trans. Signal Process.*, vol. 59, no. 12, pp. 6262–6267, Dec. 2011.

[44] H. Kwon and B. G. Lee, "Energy-efficient scheduling with delay constraints in time-varying uplink channels," *J. Commun. Networks*, vol. 10, no. 1, pp. 28–37, March 2008.

[45] G. Miao, N. Himayat, and G. Y. Li, "Energy-efficient link adaptation in frequency-selective channels," *IEEE J. Sel. Areas Commun.*, vol. 58, no. 2, pp. 545–554, Feb. 2010.

[46] C. Y. Ho and C. Y. Huang, "Non-cooperative multi-cell resource allocation and modulation adaptation for maximizing energy efficiency in uplink OFDMA cellular networks," *IEEE Wireless Commun. Lett.*, vol. 1, no. 5, pp. 420–423, Oct. 2012.

[47] M. Deruyck, D. D. Vulder, W. Joseph, and L. Martens, "Modelling the power consumption in femtocell networks," *WCNC'12 Workshop on Future Green Communications*, pp. 30–35, April 2012.

[48] R. Riggio and D. J. Leith, "A measurement-based model of energy consumption in femtocells," *Wireless Days 2012 IFIP*, pp. 1–5, Nov. 2012.

[49] A. De Domenico, E. C. Strinati, and A. Capone, "Enabling green cellular networks: a survey and outlook," *Comput. Commun.*, vol. 37, pp. 5–24, Oct. 2013.

[50] F. Meshkati, H. V. Poor, and S. C. Schwartz, "Energy-efficient resource allocation in wireless networks," *IEEE Signal Process. Mag.*, vol. 24, no. 3, pp. 58–86, May 2007.

[51] C. Isheden and G. P. Fettweis, "Energy-efficient multi-carrier link adaptation with sum rate-dependent circuit power," *Proceedings of IEEE GlobeCom'10*, pp. 1–6, Dec. 2010.

[52] Y. Rui, Q. T. Zhang, L. Deng, P. Cheng, and M. Li, "Mode selection and power optimization for energy efficiency in uplink virtual MIMO systems," *IEEE J. Sel. Areas Commun.*, vol. 31, no. 5, pp. 926–936, May 2013.

[53] G. Lim and L. J. Cimini, "Energy-efficient cooperative relaying in heterogeneous radio access networks," *IEEE Wireless Commun. Lett.*, vol. 1, no. 5, pp. 476–479, May 2013.

[54] M. Ismail, A. T. Gamage, W. Zhuang, X. Shen, E. Serpedin, and K. Qaraqe, "Uplink decentralized joint bandwidth and power allocation for energy-efficient operation in a heterogeneous wireless medium," *IEEE Trans. Commun.*, vol. 63, no. 4, pp. 1483–1495, April 2015.

[55] P. Monti, S. Tombaz, L. Wosinska, and J. Zander, "Mobile backhaul in heterogeneous network deployments: technology options and power consumption," *14th International Conference on Transparent Optical Networks*, pp. 1–7, July 2012.

[56] D. Feng, C. Jiang, G. Lim, L. J. Cimini, G. Feng, and G. Y. Li, "A survey of energy-efficient wireless communications," *IEEE Commun. Surv. Tutorials*, vol. 15, no. 1, pp. 167–178, Jan. 2013.

[57] A. P. Azad, "Analysis and optimization of sleeping mode in WiMAX via stochastic decomposition techniques," *IEEE J. Sel. Areas Commun.*, vol. 29, no. 8, pp. 1630–1640, Sept. 2011.

[58] X. Wang, A. V. Vasilakos, M. Chen, Y. Liu, and T. T. Kwon, "A survey of green mobile networks: opportunities and challenges," *Mobile Netw. Appl.*, vol. 17, no. 1, pp. 4–20, Jan. 2012.

[59] R. Mahapatra, A. De Domenico, R. Gupta, and E. C. Strinati, "Green framework for future heterogeneous wireless networks," *Mobile Netw. Appl.*, vol. 57, pp. 1518–1528, Feb. 2013.

[60] Energy Aware Radio and Network Technologies (EARTH), http://www.ict-earth.eu, Online accessed, Sept. 2012.

[61] X. Lu, E. Erkip, Y. Wang, and D. Goodman, "Power efficient multimedia communication over wireless channels," *IEEE J. Sel. Areas Commun.*, vol. 21, no. 10, pp. 1738–1751, Dec. 2003.

[62] T. H. Lan and A. H. Tewfik, "A resource management strategy in wireless multimedia communications-total power saving in mobile terminals with a guaranteed QoS," *IEEE Trans. Multimedia*, vol. 5, no. 2, pp. 267–281, Jun. 2003.

[63] T. Han and N. Ansari, "On greening cellular networks via multicell cooperation," *IEEE Wireless Commun.*, vol. 20, no. 1, pp. 82–89, Feb. 2013.

[64] E. Oh, K. Son, and B. Krishnamachari, "Dynamic base station switching-on/off strategies for green cellular networks," *IEEE Trans. Wireless Commun.*, vol. 12, no. 5, pp. 2126–2136, May 2013.

[65] C. Y. Chang, W. Liao, H. Y. Hsieh, and D. S. Shiu, "On optimal cell activation for coverage preservation in green cellular networks," *IEEE Trans. Mob. Comput.*, vol. 13, no. 11, pp. 2580–2591, Mar. 2014.

[66] N. Saxena, B. J. R. Sahu, and Y.S. Han, "Traffic-aware energy optimization in green LTE cellular systems," *IEEE Commun. Lett.*, vol. 18, no. 1, pp. 38–41, Jan. 2014.

[67] J. Wu, S. Zhou, and Z. Niu, "Traffic-aware base station sleeping control and power matching for energy-delay tradeoffs in green cellular networks," *IEEE Trans. Wireless Commun.*, vol. 12, no. 8, pp. 4196–4209, Aug. 2013.

[68] S. Navaratnarajah, A. Saeed, M. Dianati, and M. A. Imran, "Energy efficiency in heterogeneous wireless access networks," *IEEE Wireless Commun.*, vol. 20, no. 5, pp. 37–43, Oct. 2013.

[69] L. Saker, S. E. Alayoubi, R. Combes, and T. Chahed, "Optimal control of wake up mechanisms of femtocells in heterogeneous networks," *IEEE J. Sel. Areas Commun.*, vol. 30, no. 3, pp. 664–672, April 2012.

[70] A. Conte, A. Feki, L. Chiaraviglio, D. Ciullo, M. Meo, and M. A. Marsan, "Cell wilting and blossoming for energy efficiency," *IEEE Wireless Commun.*, vol. 18, no. 5, pp. 50–57, Oct. 2011.

[71] A. P. Azad, S. Alouf, E. Altman, V. Borkar, and G. S. Paschos, "Optimal control of sleep periods for wireless terminals," *IEEE J. Sel. Areas Commun.*, vol. 29, no. 8, pp. 1605–1617, Sept. 2011.

[72] A. Agarwal and A. K. Jagannatham, "Optimal wake-up scheduling for PSM delay minimization in mobile wireless networks," *IEEE Wireless Commun. Lett.*, vol. 2, no. 4, pp. 419–422, May 2013.

[73] R. Wang, J. Tsai, C. Maciocco, T. Y. C. Tai, and J. Wu, "Reducing power consumption for mobile platforms via adaptive traffic coalescing," *IEEE J. Sel. Areas Commun.*, vol. 29, no. 8, pp. 1618–1629, Sept. 2011.

[74] H. Yan, S. A. Watterson, D. K. Lowenthal, K. Li, R. Krishnan, and L. L. Peterson, "Client-centered, energy-efficient wireless communication on IEEE 802.11b networks," *IEEE Trans. Mob. Comput.*, vol. 5, no. 11, pp. 1575–1590, Nov. 2006.

[75] H. Zhu and G. Cao, "On supporting power-efficient streaming applications in wireless environments," *IEEE Trans. Mob. Comput.*, vol. 4, no. 4, pp. 391–403, Aug. 2005.

[76] Y. Jin, J. Xu, and L. Qiu, "Energy-efficient scheduling with individual packet delay constraints and non-ideal circuit power," *J. Commun. Networks*, vol. 16, no. 1, pp. 36–44, Feb. 2014.

[77] K. D. Turck, S. D. Vuyst, D. Fiems, S. Wittevrongel, and H. Bruneel, "Performance analysis of sleep mode mechanisms in the presence of bidirectional traffic," *Comput. Networks*, vol. 56, pp. 2494–2505, March 2012.

[78] L. Budzisz et al., "Dynamic resource provisioning for energy efficiency in wireless access networks: a survey and an outlook," *IEEE Commun. Surv. Tutorials*, vol. 16, no. 4, pp. 2259–2285, Sept. 2014.

[79] S. Videv, J. S. Thompson, H. Haas, and P. M. Grant, "Resource allocation for energy efficient cellular systems," *EURASIP J. Wireless Commun. Networking*, vol. 181, pp. 1–15, May 2012.

[80] Z. Ren, S. Chen, B. Hu, and W. Ma, "Energy-efficient resource allocation in downlink OFDM wireless systems with proportional rate constraints," *IEEE Trans. Veh. Technol.*, vol. 63, no. 5, pp. 2139–2150, Mar. 2014.

[81] Y. L. Chung, "Rate-and-power control based energy-saving transmissions in OFDMA-based multicarrier base stations," *IEEE Syst. J.*, vol. 9, no. 2, pp. 578–584, April 2013.

[82] L. Chen, Y. Yang, X. Chen, and G. Wei, "Energy-efficient link adaptation on Rayleigh fading channel for OSTBC MIMO system with imperfect CSIT," *IEEE Trans. Veh. Technol.*, vol. 62, no. 4, pp. 1577–1585, May 2013.

[83] L. C. Wang, W. C. Liu, A. Chen, and K. N. Yen, "Joint rate and power adaptation for wireless local area networks in generalized Nakagami Fading Channels," *IEEE Trans. Veh. Technol.*, vol. 58, no. 3, pp. 1375–1386, March 2009.

[84] L. P. Qian, Y. J. Zhang, Y. Wu, and J. Chen, "Joint base station association and power control via benders' decomposition," *IEEE Trans. Wireless Commun.*, vol. 12, no. 4, pp. 1651–1665, April 2013.

[85] V. A. Siris and M. Anagnostopoulou, "Performance and energy efficiency of mobile data offloading with mobility prediction and prefetching," *Proceedings of IEEE WoWMoM'13*, pp. 1–6, June 2013.

[86] N. Ristanovic, J. Y. Le Boudec, A. Chaintreau, and V. Erramilli, "Energy efficient offloading of 3G networks," *Proceedings of IEEE MASS'11*, pp. 202–211, Oct. 2011.

[87] S. Tarkoma, M. Siekkinen, E. Lagerspetz, and Y. Xiao, *Smartphone Energy Consumption: Modeling and Optimization*, Cambridge University Press, 2014.

[88] T. A. Le, S. Nasseri, A. Z. Esfahani, M. R. Nakhai, and A. Mills, "Power-efficient downlink transmission in multicell networks with limited wireless backhaul," *IEEE Wireless Commun.*, vol. 18, no. 5, pp. 82–88, Oct. 2011.

[89] M. Z. Shakir, K. A. Qaraqe, H. Tabassum, M. S. Alouini, E. Serpedin, and M. A. Imran, "Green heterogeneous small-cell networks: toward reducing the CO_2 emissions of mobile communications industry using uplink power adaptation," *IEEE Commun. Mag.*, vol. 51, no. 6, pp. 52–61, Jun. 2013.

[90] T. Elkourdi and O. Simeone, "Femtocell as a relay: an outage analysis," *IEEE Trans. Wireless Commun.*, vol. 10, no. 12, pp. 4204–4213, Oct. 2011.

[91] F. Parzysz, M. Vu, and F. Gagnon, "Impact of propagation environment on energy-efficient relay placement: model and performance analysis," *IEEE Trans. Wireless Commun.*, vol. 13, no. 4, pp. 2214–2228, April 2014.

[92] Y. Li, X. Zhu, C. Liao, C. Wang, and B. Cao, "Energy efficiency maximization by jointly optimizing the positions and serving range of relay stations in cellular networks," *IEEE Trans. Veh. Technol.*, vol. 64, no. 6, pp. 2551–2560, July 2014.

[93] M. F. Uddin, C. Assi, and A. Ghrayeb, "Joint relay assignment and power allocation for multicast cooperative networks," *IEEE Commun. Lett.*, vol. 16, no. 3, pp. 368–371, March 2012.

[94] K. Cheung, S. Yang, and L. Hanzo, "Achieving maximum energy-efficiency in multi-relay OFDMA cellular networks: a fractional programming approach," *IEEE Trans. Commun.*, vol. 61, no. 7, pp. 2746–2757, July 2013.

[95] R. A. Loodaricheh, S. Mallick, and V. K. Bhargava, "Energy-efficient resource allocation for OFDMA cellular networks with user cooperation and QoS provisioning," *IEEE Trans. Wireless Commun.*, vol. 13, no. 11, pp. 6132–6146, Nov. 2014.

[96] Y. Li, C. Liao, and C. Wang, "Energy-efficient optimal relay selection in cooperative cellular networks based on double auction," *IEEE Trans. Wireless Commun.*, vol. 14, no. 8, pp. 4093–4014, March 2015.

[97] G. Zhang, "Subcarrier and bit allocation for real-time services in multiuser OFDM systems," *Proceedings of IEEE ICC04*, pp. 2985–2989, June 2004.

[98] G. Youjun, T. Hui, Z. Ping, and X. Haibo, "A QoS-Guaranteed adaptive resource allocation algorithm with low complexity in OFDMA system," *Proceedings of International Conference on Wireless Communications Networking and Mobile Computing*, pp. 1–4, Sept. 2006.

[99] M. Jung, K. Hwang, and S. Choi, "Joint mode selection and power allocation scheme for power-efficient Device-to-Device (D2D) communication," *Proceedings of the IEEE VTC-Spring '12*, pp. 1–5, May 2012.

[100] M. Belleschi, G. Fodor, and A. Abrardo, "Performance analysis of a distributed resource allocation scheme for D2D communications," *Proceedings of IEEE GLOBECOM Workshops*, pp. 358–362, Dec. 2011.

[101] M. R. Gary and D. S. Johnson, *Computers and Intractability: A Guide to the Theory of NP-Completeness*, Freeman, San Francisco, CA, 1979.

[102] X. Xiao, X. Tao, and J. Lu, "A QoS-aware power optimization scheme in OFDMA systems with integrated Device-to-Device (D2D) communications," *Proceedings of IEEE VTC-Fall'11*, pp. 1–5, Sept. 2011.

[103] G. Fodor et al., "Design aspects of network assisted device-to-device communications," *EEE Commun. Mag.*, vol. 50, no. 3, pp. 170–177, March 2012.

[104] A. Asadi and V. Mancuso, "Energy efficient opportunistic uplink packet forwarding in hybrid wireless networks," *Proceedings of the 4th International Conference on Future Energy Systems*, pp. 261–262, 2013.

[105] H. Ghazzai, E. Yaacoub, M. S. Alouini, and A. A. Dayya, "Optimized smart grid energy procurement for LTE networks using evolutionary algorithms," *IEEE Trans. Veh. Technol.*, vol. 63, no. 9, pp. 4508–4519, March 2014.

[106] C. McGuire, M. R. Brew, F. Darbari, G. Bolton, A. McMahon, D. H. Crawford, S. Weiss, and R. W. Stewart, "HopScotch-a low-power renewable energy base station network for rural broadband access," *EURASIP J. Wireless Commun. Networking*, vol. 112, pp. 1–12, March 2012.

[107] T. Han and N. Ansari, "Powering mobile networks with green energy," *IEEE Wireless Commun.*, vol. 21, no. 1, pp. 90–96, Feb. 2014.

[108] B. Devillers and D. Gunduz, "A general framework for the optimization of energy harvesting communication systems with battery imperfections," *J. Commun. Networks*, vol. 14, no. 2, pp. 130–139, April 2012.

[109] O. Orhan, D. Gunduz, and E. Erkip, "Throughput maximization for an energy harvesting communication system with processing cost," *Proceedings of IEEE Information Theory Workshop*, pp. 84–88, Sept. 2012.

[110] K. Tutuncuoglu and A. Yener, "Optimum transmission policies for battery limited energy harvesting nodes," *IEEE Trans. Wireless Commun.*, vol. 11, no. 3, pp. 1180–1189, Feb. 2012.

[111] D. W. K. Ng, E. S. Lo, and R. Schober, "Energy-efficient resource allocation in OFDMA systems with hybrid energy harvesting base station," *IEEE Trans. Wireless Commun.*, vol. 12, no. 7, pp. 3412–3427, July 2013.

[112] K. Tutuncuoglu and A. Yener, "Sum-rate optimal power policies for energy harvesting transmitters in an interference channel," *J. Commun. Networks*, vol. 14, no. 2, pp. 151–161, April 2012.

[113] O. Ozel, K. Tutuncuoglu, J. Yang, S. Ulukus, and A. Yener, "Transmission with energy harvesting nodes in fading wireless channels: optimal policies," *IEEE J. Sel. Areas Commun.*, vol. 29, no. 8, pp. 1732–1743, Sept. 2011.

[114] B. Gurakan, O. Ozel, J. Yang, and S. Ulukus, "Energy cooperation in energy harvesting communications," *IEEE Trans. Commun.*, Available on: arXiv:1303.2636v1 [cs.IT] 11 March 2013.

[115] C. F. Chiasserini and R. R. Rao, "Energy efficient battery management," *IEEE J. Sel. Areas Commun.*, vol. 19, no. 7, pp. 1235–1245, July 2001.

[116] A. Asadi, Q. Wang, and V. Mancuso, "A survey on device-to-device communication in cellular networks," *IEEE Commun. Surv. Tutorials*, vol. 16, no. 4, pp. 1801–1819, Sept. 2014.

[117] https://s3-us-west-2.amazonaws.com/belllabs-microsite-greentouch/index.php@page=about-us .html.

[118] https://www.ict-earth.eu/.

[119] http://www.celtic-initiative.org/Projects/OPERA-Net, Online accessed, April 2016.

[120] http://www.meti.go.jp/english/policy/GreenITInitiativeJapan.pdf, Online accessed, April 2016.

[121] M. C. Erturk, S. Mukherjee, H. Ishii, and H. Arslan, "Distributions of transmit power and SINR in device-to-device networks," *IEEE Commun. Lett.*, vol. 17, no. 2, pp. 273–276, Feb. 2013.

[122] B. Zhou, H. Hu, S.-Q. Huang, and H.-H. Chen, "Intracluster device-to-device relay algorithm with optimal resource utilization," *IEEE Trans. Veh. Technol.*, vol. 62, no. 5, pp. 2315–2326, Jun. 2013.

[123] T. Kim and M. Dong, "An iterative hungarian method to joint relay selection and resource allocation for D2D communications," *IEEE Wireless Commun. Lett.*, vol. 3, no. 6, pp. 625–628, Dec. 2014.

[124] H. Cai, I. Koprulu, and N. Shroff, "Exploiting double opportunities for deadline based content propagation in wireless networks," *Proceedings of IEEE INFOCOM'13*, pp. 764–772, 2013.

[125] N. Golrezaei, A. F. Molisch, and A. G. Dimakis, "Base-station assisted device-to-device communications for high-throughput wireless video networks," *Proceedings of IEEE ICC'12*, pp. 7077–7081, 2012.

[126] M. Ismail and W. Zhuang, "Green radio communications in a heterogeneous wireless medium," *IEEE Wireless Commun. Mag.*, vol. 21, no. 3, pp. 128–135, Jun. 2014.

[127] M. F. Marzban, M. Ismail, M. Abdallah, M. Khairy, K. Qaraqe, and E. Serpedin, "IDC interference-aware resource allocation for LTE/WLAN heterogeneous networks," *IEEE Wireless Commun. Lett.*, vol. 4, no. 6, pp. 581–584, Aug. 2015.

[128] 3GPP, "TR 36.816: Evolved Universal Terrestrial Radio Access (E-UTRA); Study on Signaling and Procedure for Interference Avoidance for in-Device Coexistence," 2012.

[129] Z. Hu, R. Susitaival, Z. Chen, I. Fu, P. Dayal, and S. Baghel, "Interference avoidance for in-device coexistence in 3GPP LTE-advanced: challenges and solutions," *IEEE Commun. Mag.*, vol. 50, no. 11, pp. 60–67, Nov. 2012.

[130] J. Choi, J. Yoo, S. Choi, and C. Kim, "EBA: an enhancement of the IEEE 802.11 DCF via distributed reservation," *IEEE Trans. Mob. Comput.*, vol. 4, no. 4, pp. 378–390, Aug. 2005.

[131] IEEE Std 802.11, "Part 11: Wireless LAN Medium Access Control (MAC) and Physical Layer (PHY) Specifications," 2012.

[132] A. R. Ekti, X. Wang, M. Ismail, E. Serpedin, and K. Qaraqe, "Joint user association and data rate allocation in heterogeneous wireless networks," *IEEE Trans. Veh. Technol.*, IEEE Early Access since Nov. 2015.

[133] M. Ismail, E. Serpedin, and K. Qaraqe, "A win-win cooperative downlink resource allocation for green communications in a heterogeneous wireless medium," *2nd Workshop on Green Broadband Access - IEEE Globecom'14*, Dec. 2014.

[134] H. Lee, S. Vahid, and K. Moessner, "A survey of radio resource management for spectrum aggregation in LTE-advanced," *IEEE Wireless Commun. Surv. Tutorials*, vol. 16, no. 2, pp. 745–760, May 2014.

[135] M. B. Celebi, I. Guvenc, and H. Arslan, "Interference mitigation for LTE uplink through iterative blanking," *IEEE Globecom'11*, pp. 1–6, Dec. 2011.

[136] B. Gedik, O. Amin, and M. Uysal, "Power allocation for cooperative systems with training-aided channel estimation," *IEEE Trans. Wireless Commun.*, vol. 8, no. 9, pp. 4773–4783, Sept. 2009.

[137] M. Ismail, K. Qaraqe, and E. Serpedin, "Cooperation incentives and downlink radio resource allocation for green communications in a heterogeneous wireless environment," *IEEE Trans. Veh. Technol.*, vol. 65, no. 3, pp. 1627–1638, March 2015.

[138] D. L. Perez, A. Valcarce, G. de la Roche, and J. Zhang, "OFDMA Femtocells: a roadmap on interference avoidance," *IEEE Commun. Mag.*, vol. 47, no. 9, pp. 41–48, Sept. 2009.

[139] H. C. Lee, D. C. Oh, and Y. H. Lee, "Mitigation of interfemtocell interference with adaptive fractional frequency reuse," *Proceedings of IEEE ICC'10*, pp. 1–5, May 2010.

[140] A. Mahmud and K. A. Hamdi, "A unified framework for the analysis of fractional frequency reuse techniques," *IEEE Trans. Commun.*, vol. 62, no. 10, pp. 3692–3705, Oct. 2014.

[141] V. Chandrasekhar and J. G. Andrews, "Spectrum allocation in tiered cellular networks," *IEEE Trans. Commun.*, vol. 57, no. 10, pp. 3059–3068, Oct. 2009.

[142] S. Boyd and L. Vandenberghe, *Convex Optimization*, Cambridge University Press, 2009.

[143] J. Nash, "The bargaining problem," *Econometrica*, vol. 18, no. 2, pp. 155–162, April 1950.

[144] H. Boche, M. Schubert, N. Vucic, and S. Naik, "Non-symmetric nash bargaining solution for resource allocation in wireless networks and connection to interference calculus," *Proceedings of the 15th European Signal Processing Conference*, 2007.

[145] H. Yaiche, R. R. Mazumdar, and C. Rosenberg, "A game theoretic framework for bandwidth allocation and pricing in broadband networks," *IEEE/ACM Trans. Networking*, vol. 8, no. 5, pp. 667–678, Oct. 2000.

[146] C. Liu, X. Qin, S. Zhang, and W. Zhou, "Proportional-fair downlink resource allocation in OFDMA-based relay networks," *J. Commun. Networks*, vol. 13, no. 6, pp. 633–638, Dec. 2011.

[147] C. Xiong, G. Y. Li, S. Zhang, Y. Chen, and S. Xu, "Energy- and spectral-efficiency tradeoff in downlink OFDMA networks," *IEEE Trans. Wireless Commun.*, vol. 10, no. 11, pp. 3874–3886, Nov. 2011.

[148] C. Xiong, G. Y. Li, S. Zhang, Y. Chen, and S. Xu, "Energy-efficient resource allocation in OFDMA networks," *IEEE Trans. Commun.*, vol. 60, no. 12, pp. 3767–3778, Dec. 2012.

[149] D. W. K. Ng, E. S. Lo, and R. Schober, "Energy-efficient resource allocation in OFDMA systems with large numbers of base station antennas," *IEEE Trans. Wireless Commun.*, vol. 11, no. 9, pp. 3292–3304, Sept. 2012.

[150] T. S. Chang, K. T. Feng, J. S. Lin, and L. C. Wang, "Green resource allocation schemes for relay-enhanced MIMO-OFDM networks," *IEEE Trans. Veh. Technol.*, vol. 62, no. 9, pp. 4539–4554, Nov. 2013.

[151] W. C. Chung, C. J. Chang, and C. Y. Huang, "A green radio resource allocation scheme for LTE-A CoMP systems with multimedia traffic," *Proceedings of IEEE/CIC ICCC'13*, pp. 758–762, Aug. 2013.

[152] O. Arnold, F. Richter, G. Fettweis, and O. Blume, "Power consumption modeling of different base station types in heterogeneous cellular networks," *Proceedings of ICT-MobileSummit*, June 2010.

[153] S. Kim, B. G. Lee, and D. Park, "Radio resource allocation for energy consumption minimization in multi-homed wireless networks," *Proceedings of IEEE ICC13*, pp. 5589–5594, June 2013.

[154] M. Ismail, A. T. Gamage, W. Zhuang, and X. Shen, "Energy efficient uplink resource allocation in a heterogeneous wireless medium," *Proceedings of IEEE ICC 2014*, pp. 5275–5280, June 2014.

[155] L. Golubchik, J. C. S. Lui, T. F. Tung, A. L. H. Chow, W. J. Lee, G. Franceschinis, and C. Anglano, "Multi-path continuous media streaming: what are the benefits?" *Perform. Eval.*, vol. 49, no. 1, pp. 429–449, Sept. 2002.

[156] M. D. Trott, "Path diversity for enhanced media streaming," *IEEE Commun. Mag.*, vol. 42, no. 8, pp. 80–87, Aug. 2004.

[157] G. P. Perrucci, F. H.P. Fitzek, and J. Widmer, "Survey on energy consumption entities on the smartphone platform," *Proceedings of IEEE VTC 2011*, pp. 1–6, May 2011.

[158] L. Lei, Z. Zhong, C. Lin, and X. Shen, "Operator controlled device-to-device communications in LTE-advanced networks," *IEEE Wireless Commun.*, vol. 19, no. 3, pp. 96–104, Jun. 2012.

[159] E. Rantalai, A. Karppanen, S. Granlund, and P. Sarolahti, "Modeling energy efficiency in wireless internet communication," *Proceedings of MobiHeld '09. ACM*, pp. 67–72, Aug. 2009.

[160] R. Litjens, H. van den Berg, and R. J. Boucherie, "Throughputs in processor sharing models for integrated stream and elastic traffic," *Proceedings of MobiHeld '09. ACM*, vol. 65, no. 2, pp. 152–180, Feb. 2008.

[161] M. Ismail and W. Zhuang, "Mobile terminal energy management for sustainable multi-homing video transmission," *IEEE Trans. Wireless Commun.*, vol. 13, no. 8, pp. 4616–4626, Aug. 2014.

[162] N. Abu-Ali, A. M. Taha, M. Salah, and H. Hassanein, "Uplink scheduling in LTE and LTE-advanced: tutorial, survey, and evaluation framework," *IEEE Wireless Commun. Surv. Tutorials*, vol. 16, no. 3, pp. 1239–1265, Aug. 2014.

[163] F. Liu, K. Zheng, W. Xiang, and H. Zhao, "Design and performance analysis of an uplink carrier aggregation scheme," *IEEE J. Sel. Top. Signal Process.*, vol. 32, no. 2, pp. 197–207, Feb. 2014.

[164] J. B. G. Frenk and S. Schaible, "Fractional programming," *ERIM Report Series Research in Management*, ERS-2004074-LIS, p. 55, 2004.

[165] J. P. G. Crouzeix and J. A. Ferland, "Algorithms for generalized fractional programming," *Math. Program.*, vol. 52, no. 13, pp. 191–207, May. 1991.

[166] Technical Specification, "LTE; Evolved universal terrestrial radio access (E-UTRA); User equipment (UE) radio transmission and reception - (3GPP TS 36.101 version 10.3.0 Release 10)," June 2011.

[167] Technical Specification, "Universal Mobile Telecommunications System (UMTS); User Equipment (UE) radio transmission and reception (FDD) - (3GPP TS 25.101 version 6.19.0 Release 6)," March 2009.

[168] K. Pandit, A. Ghosh, D. Ghosal, and M. Chiang, "Content aware optimization for video delivery over WCDMA," *EURASIP J. Wireless Commun. Networking*, vol. 2012, no. 1, p. 217, July 2012.

[169] J. Chakareski, S. Han, and B. Girod, "Layered coding vs. multiple description for video streaming over multiple paths," *Proceedings of ACM International Conference on Multimedia*, pp. 422–431, 2003.

[170] F. Fu and M. van der Schaar, "Structural solutions for dynamic scheduling in wireless multimedia transmission," *IEEE Trans. Circuits Syst. Video Technol.*, vol. 22, no. 5, pp. 727–739, May 2012.

[171] D. Jurca and P. Frossard, "Video packet selection and scheduling for multipath streaming," *IEEE Trans. Multimedia*, vol. 9, no. 3, pp. 629–641, April 2007.

[172] D. Fiems, B. Steyaert, and H. Bruneel, "A genetic approach to Markovian characterisation of H.264 scalable video," *Multimedia Tools Appl.*, vol. 58, no. 1, pp. 125–146, May 2012.

[173] M. van der Schaar and D. Turaga, "Cross-layer packetization and retransmission strategies for delay-sensitive wireless multimedia transmission," *IEEE Trans. Multimedia*, vol. 9, no. 1, pp. 185–197, Jan. 2007.

[174] Y. Zhang, F. Fu, and M. van der Schaar, "On-line learning and optimization for wireless video transmission," *IEEE Trans. Signal Process.*, vol. 58, no. 6, pp. 3108–3124, Jun. 2010.

[175] H. Kellerer, U. Pferschy, and D. Pisinger, *Knapsack Problems*, Springer-Verlag, 2004.

[176] S. Martello and P. Toth, "Heuristic algorithms for the multiple knapsack problem," *Computing*, vol. 27, no. 2, pp. 93–112, 1981.

[177] K. F. Man, K. S. Tang, and S. Kwong, "Genetic algorithms: concepts and applications," *IEEE Trans. Ind. Electron.*, vol. 43, no. 5, pp. 519–534, Oct. 1996.

[178] M. V. Bhalerao, S. S. Sonavane, and V. Kumar, "A survey of wireless communications using visible light," International Journal of Advances in Engineering & Technology, pp. 1–10, Jan. 2013.

[179] D. Gujjari, "Visible light communication," *MSc. Thesis*, Dalhousie University, Aug. 2012.

[180] R. Zhang, J. Wang, Z. Wang, Z. Y. Xu, C. Zhao, and L. Hanzo, "Visible light communications in heterogeneous networks: paving the way for user-centric design," *IEEE Wireless Commun.*, vol. 22, no. 2, pp. 8–16, April 2015.

[181] M. Saadi, L. Wattisuttikulkij, Y. Zhao, and P. Sangwongngam, "Visible light communication: opportunities, challenges, and channel models," *Int. J. Electron. Inf.*, vol. 2, no. 1, pp. 1–11, Feb. 2013.

[182] S. Shao, A. Khreishah, M. B. Rahaim, and H. Elgala, "An indoor hybrid WiFi-VLC internet access system," *Proceedings of the IEEE 11th International Conference on Mobile Adhoc and Sensor Systems*, pp. 569–574, 2014.

[183] M. Kashef, M. Ismail, M. Abdallah, K. Qaraqe, and E. Serpedin, "Energy efficient resource allocation for mixed RF/VLC heterogeneous wireless networks," *IEEE J. Sel. Areas Commun.*, IEEE Early Access with date: March 2016.

[184] S. Rajagopal, R. D. Roberts, and S. K. Lim, "IEEE 802.15.7 visible light communication: modulation schemes and dimming support," *IEEE Commun. Mag.*, vol. 50, no. 3, pp. 72–82, March 2012.

[185] I. Stefan and H. Haas, "Hybrid visible light and radio frequency communication systems," *IEEE VTC, Fall14*, pp. 1–5, Sept. 2014.

[186] J. H. Liu, Q. Li, and X. Y. Zhang, "Cellular coverage optimization for indoor visible light communication and illumination networks," *J. Commun.*, vol. 9, no. 11, pp. 891–898, Nov. 2014.

[187] C. Chen, D. Tsonev, and H. Haas, "Joint transmission in indoor visible light communication downlink cellular networks," *Proceedings of IEEE Globecom Workshop*, pp. 1127–1132, Dec. 2013.

[188] X. Li, R. Zhang, and L. Hanzo, "Cooperative load balancing in hybrid visible light communications and WiFi," *IEEE Trans. Commun.*, vol. 63, no. 4, pp. 1319–1329, April 2015.

[189] H. Chowdhury, I. Ashraf, and M. Katz, "Energy-efficient connectivity in hybrid radio-optical wireless systems," *Proceedings of the 10th International Symposium on Wireless Communications Systems*, pp. 1–5, Aug. 2013.

[190] D. A. Basnayaka and H. Haas, "Hybrid RF and VLC systems: improving user data rate performance of VLC systems," *Proceedings of IEEE VTC*, Spring, to appear.

[191] F. Jin, R. Zhang, and L. Hanzo, "Resource allocation under delay-guarantee constraints for heterogeneous visible light and RF femtocell," *IEEE Trans. Wireless Commun.*, vol. 14, no. 2, pp. 1020–1034, Feb. 2015.

[192] D. Bykhovsky and S. Arnon, "Multiple access resource allocation in visible light communication systems," *J. Lightwave Technol.*, vol. 32, no. 8, pp. 1594–1600, March 2014.

[193] M. Kashef, M. Abdallah, K. Qaraqe, H. Haas, and M. Uysal, "On the benefits of cooperation via power control in OFDM-based visible light communication systems," *IEEE PIMRC*, 2014.

[194] I. Stefan and H. Haas, "Analysis of optimal placement of LED arrays for visible light communication," *2013 IEEE VTC*, Spring, pp. 1–5, 2–5 June 2013.

[195] L. Saker, S.-E. Elayoubi, and T. Chahed, "Minimizing energy consumption via sleep mode in green base station," *Proceedings of IEEE WCNC*, pp. 1–6, April 2010.

[196] H. Y. Lateef, M. Z. Shakir, M. Ismail, A. Mohamed, and K. Qaraqe, "Towards energy efficient and quality of service aware cell zooming in 5G wireless networks," *Proceedings of IEEE VTC*, to appear.

[197] M. Ismail, M. Kashef, E. Serpedin, and K. Qaraqe, "On balancing energy efficiency for network operators and mobile users in dynamic planning," *IEEE Commun. Mag.*, vol. 53, no. 11, pp. 158–165, Nov. 2015.

[198] M. Ismail, M. Kashef, E. Serpedin, and K. Qaraqe, "Dynamic planning with balanced energy efficiency for network operators and mobile users," *IEEE OnlineGreenComm*, pp. 1–6, Nov. 2014.

[199] X. Li, H. Wang, N. Liu, and X. You, "Dynamic user association for energy minimization in macro-relay network," *Proceedings of IEEE WCSP*, pp. 1–5, 2012.

[200] X. Li, H. Wang, C. Meng, X. Wang, and N. Liu, "Total energy minimization through dynamic station-user connection in macro-relay network," *Proceedings of IEEE WCNC*, pp. 697–702, 2013.

[201] R. Fantini, D. Sabella, and M. Caretti, "Energy efficiency in LTEAdvanced networks with relay nodes," *Proceedings of IEEE VTC*, pp. 1–5, 2011.

[202] I. Stefan, H. Burchardt, and H. Haas, "Area spectral efficiency performance comparison between VLC and RF femtocell networks," *IEEE ICC'13*, pp. 3825–3829, June 2013.

[203] V. Goswami and U. C. Gupta, "Analyzing the discrete-time multiserver queue Geom/Geom/m queue with late and early arrivals," *Inf. Manage. Sci.*, vol. 9, no. 2, pp. 55–66, Jun. 1998.

[204] Y. Chen, S. Zhang, and S. Xu, "Characterizing energy efficiency and deployment efficiency relations for green architecture design," *Proceedings of IEEE International Conference on Communications, (ICC'2010)*, pp. 1–5, June 2010.

[205] 3rd Generation Partnership Project (3GPP)–Technical Specification Group Services and System Aspects (TSG SA) Telecommunication Management, "Study on energy savings management (ESM) (Release 10)," *3GPP Technical Report TR 32.826 V10.0.0 (2010–03)*, 3GPP, France, pp. 1–33, Mar. 2010.

[206] L. M. Correia, D. Zeller, O. Blume, et al., "Challenges and enabling technologies for energy aware mobile radio networks," *IEEE Commun. Mag.*, vol. 48, no. 11, pp. 66–72, Nov. 2010.

[207] X. Wang, A. V. Vasilakos, M. Chen, Y. Liu, and T. T. Kwon, "A survey of green mobile networks: opportunities and challenges," *Springer J. Mob. Networks Appl.*, vol. 17, no. 1, pp. 4–20, 2012.

[208] P. Lin, J. Zhang, Y. Chen, and Q. Zhang, "Macro-femto heterogeneous network deployment and management: from business models to technical solutions," *IEEE Wireless Commun. Mag.*, vol. 18. no. 3, pp. 64–70, Jun. 2011.

[209] S. Landstrom, A. Furuskar, K. Johansson, L. Falconetti, and F. Kronestedt, "Heterogeneous networks increasing cellular capacity," *J. Ericson Rev.*, vol. 89, pp. 4–9, Jan. 2011.

[210] N. Shetty, S. Parekh, and J. Walrand, "Economics of femtocells," *Proceedings of IEEE Conference on Global Communications, (GLOBECOM'09)*, pp. 1–5, Dec. 2009.

[211] H. Claussen, L. T. W. Ho, and L. G. Samuel, "An overview of the femtocell concept," *Bell Labs Tech. J.*, vol. 13, no. 1, pp. 221–245, May 2008.

[212] V. Chandrasekhar, J. Andrews, and A. Gatherer, "Femtocell networks: a survey," *IEEE Commun. Mag.*, vol. 46, no. 9, pp. 59–67, Sept. 2008.

[213] I. Guvenc, "Capacity and fairness analysis of heterogeneous networks with range expansion and interference coordination," *IEEE Commun. Lett.*, vol. 15, no. 10, pp. 1084–1087, Oct. 2011.

[214] M. C. Erturk, I. Guvenc, S. Mukherjee, and H. Arslan, "Fair and QoS-oriented resource management in heterogeneous networks," *EURASIP J. Wireless Commun. Networking*, vol. 2013, no. 1, p. 121, May 2013.

[215] D. Calin, H. Claussen, and H. Uzunalioglu, "On femto deployment architectures and macrocell offloading benefits in joint macro-femto deployments," *IEEE Commun. Mag.*, vol. 48, no. 1, pp. 26–32, Jan. 010.

[216] M. Bennis, D. Niyato, and T. Alpcan, Distributed Learning Strategies for Femtocell Networks, *Femtocell Networks: Deployment, PHY techniques, and Resource Management*, Cambridge Press, UK, Apr. 2013.

[217] J. Hoydis, M. Kobayashi, and M. Debbah, "Green small-cell networks," *IEEE Mag. Veh. Technol.*, vol. 6, no. 1, pp. 37–43, Mar. 2011.

[218] F. Richter, A. J. Fehske, and G. P. Fettweis, "Energy efficiency aspects of base station deployment strategies for cellular networks," *IEEE 70th Vehicular Technology Conference, (VTC-Fall'2009)*, pp. 1–5, Sep. 2009.

[219] A. J. Fehske, F. Richter, and G. P. Fettweis, "Energy efficiency improvements through micro sites in cellular mobile radio networks," *Proceedings of IEEE Conference on Global Communications Workshops, (GLOBECOM'2009)*, pp. 1–5, Dec. 2009.

[220] H. Claussen, L. T. W. Ho, and F. Pivit, "Effects of joint macrocell and residential picocell deployment on the network energy efficiency," *Proceedings of IEEE 19th International Symposium on Personal, Indoor and Mobile Radio Communications, (PIMRC'2008)*, pp. 1–6, Sept. 2008.

[221] A. R. Ekti, M. Z. Shakir, E. Serpedin, and K. A. Qaraqe, "Characterizing energy efficiency and deployment efficiency relations for green architecture design," *Proceedings of IEEE International Conference on Communications, (ICC'2010)*, pp. 1–5, Cape Town, South Africa, June 2010.

[222] M.-S. Alouini and A. Goldsmith, "Area spectral efficiency of cellular mobile radio systems," *IEEE Trans. Veh. Technol.*, vol. 48, no. 4, pp. 1047–1066, July 1999.

[223] S. R. Saunders and A. A. Zavala, *Antenna and Propagation for Wireless Communication Systems*, 2nd ed., John Wiley & Sons, Ltd., Chicester, UK, Mar. 2007.

[224] P. Harley, "Short distance attenuation measurements at 900 MHz and 1.8 GHz using low antenna heights for microcells," *IEEE. J. Sel. Areas Commun.*, vol. SAC-7, pp. 5–11, Jan. 1989.

[225] 3rd Generation Partnership Project (3GPP)–Technical Specification Group Radio Access Network (TSG RAN) Radio Layer 1, "Physical layer aspects for evolved universal terrestrial radio access (UTRA) (Release 9)," *3GPP Technical Report TR 25.814 V7.1.0 (2006–09)*, pp. 1–132, Oct. 2006.

[226] A. Simonsson and A. Furuskar, "Uplink power control in LTE -overview and performance, subtitle: principles and benefits of utilizing rather than compensating for SINR variations," *Proceedings of IEEE 68th Vehicular Technology Conference, (VTC-Fall'2008)*, pp. 1–5, Calgary, AB, Canada, Sept. 2008.

[227] B. Muhammad and A. Mohammed, "Performance evaluation of uplink closed loop power control for LTE system," *Proceedings of IEEE 70th Vehicular Technology Conference, (VTC-Fall'2009)*, pp. 1–5, Anchorage, AK, USA, Sept. 2009.

[228] A. M. Rao, "Reverse link power control for managing inter-cell interference in orthogonal multiple access systems," *Proceedings of IEEE 66th Vehicular Technology Conference, (VTC-Fall'2007)*, pp. 1–5, Baltimore, MD, USA, Oct. 2007.

[229] S. Al-Ahmadi and H. Yanikomeroglu, "On the approximation of the generalized-K PDF by a Gamma PDF using the moment matching method," *Proceedings of IEEE Conference on Wireless Communications and Networking, (WCNC'09)*, pp. 1–6, Budapest, Hungary, April 2009.

[230] R. K. Mallik, "A new statistical model of the complex Nakagami-m fading gain," *IEEE Trans. Commun.*, vol. 58, no. 9, pp. 2611–2620, 2010.

[231] P. Bithas, N. Sagias, P. Mathiopoulos, G. Karagiannidis, and A. Rontogiannis, "On the performance analysis of digital communications over generalized-K fading channels," *IEEE Commun. Lett.*, vol. 5, no. 10, pp. 353–355, May 2006.

[232] Y. Kim, T. Kwon, and D. Hong, "Area spectral efficiency of shared spectrum hierarchical cell structure networks," *IEEE Trans. Veh. Technol.*, vol. 59, no. 8, pp. 4145–4151, 2010.

[233] K. A. Hamdi, "A useful lemma for capacity analysis of fading interference channels," *IEEE Trans. Commun.*, vol. 58, no. 2, pp. 411–416, Feb. 2010.

[234] T. Persson, C. Trnevik, L.-E. Larsson, and J. Lovn, "Output power distributions of terminals in a 3G mobile communication network," *Wiley J. Bioelectromagn.*, vol. 33, no. 4, pp. 320–325, May 2012.

[235] H. Tabassum, M. Z. Shakir, and M. Alouini, "On the area green efficiency (AGE) of heterogeneous networks," *Proceedings of International Conference on Global Communications, GLOBE-COM'2012*, pp. 1–6, Anaheim, CA, USA, Dec. 2012.

[236] Energy and carbon conversions: fact sheet, http://www.carbontrust.com, Online accessed, Nov. 2012.

[237] International Monetary Fund, World Economic Outlook (WEO) Database Groups and Aggregates Information, http://www.imf.org/external/ pubs/ft/weo/2012/01/weodata/ groups.htm, Online accessed, April 2012.

[238] S. Gradshteyn and I. M. Ryzhik, *Table of Integrals, Series, and Products*, 6th ed., Academic Press, New York, 2000.

[239] UMTS Forum, UMTS Forum Report: Mobile Traffic Forecasts: 2010-2020, Jan. 2011.

[240] Ericsson, More Than 50 Billion Connected Devices, White paper, Feb. 2011.

[241] H. Ishii, Y. Kishiyama, and H. Takahashi, "A novel architecture for LTE-B: C-plane/U-plane split and Phantom Cell concept," *2012 IEEE Globecom Workshops (GC Wkshps)*, pp. 624–630, Dec. 2012.

[242] J. G. Andrews, H. Glaussen, M. Dohler, S. Rangan, and M. C. Reed, "Femtocells: past, present, and future," *IEEE J. Sel. Areas Commun.*, vol. 30, no. 3, pp. 497–508, April 2012.

[243] A. Damnjanovic, J. Montojo, W. Yongbin, J. Tingfang, L. Tao, M. Vajapeyam, Y. Taesang, S. Osok, and D. Malladi, "A survey on 3GPP heterogeneous networks," *IEEE Wireless Commun. Mag.*, vol. 18, no. 3, pp. 10–21, Jun. 2011.

[244] I. Guvenc, "Capacity and fairness analysis of heterogeneous networks with range expansion and interference coordination," *IEEE Commun. Lett.*, vol. 15, no. 10, pp. 1084–1087, 2011.

[245] K. Okino, T. Nakayama, C. Yamazaki, H. Sato, and Y. Kusano, "Pico cell range expansion with interference mitigation toward LTE-Advanced heterogeneous networks," *Proceedings of IEEE International Workshop on Heterogeneous Networks*, pp. 1–5, June 2011.

[246] I. Shgluof, M. Ismail, and R. Nordin, "Efficient femtocell deployment under macrocell coverage in LTE-Advanced system," *Proceedings of the International Conference on Computing, Management and Telecommunications (ComManTel)*, pp. 60–65, Jan. 2013.

[247] M. Z. Shakir and M.-S. Alouini, "On the area spectral efficiency improvement of heterogeneous network by exploiting the integration of macro-femto cellular networks," *Proceedings of IEEE International Conference on Communications*, pp. 1–6, June 2012.

[248] Informa Telecoms & Media, "Small cell market status," *Report for Small Cell Forum (Issue 2)*, pp. 1–14, June 2012.

[249] M. Z. Shakir, H. Tabassum, K. Qaraqe, E. Serpedin, M.-S. Alouini, and M. A. Imran, "Green heterogeneous small-cell networks: toward reducing the CO2 emissions of mobile communication industry via uplink adaptation," *IEEE Commun. Mag.*, vol. 51, no. 6, pp. 52–61, Jun. 2013.

[250] O. Tipmongkolsilp, S. Zaghloul, and A. Jukan, "The evolution of cellular backhaul technologies: current issues and future trends," *IEEE Commun. Surv. Tutorials*, vol. 13, no. 1, pp. 97–113, 2011.

[251] M. Paolini, L. Hiley, and F. Rayal, Small-cell backhaul: industry trends and market overview, http://www.proxim.com/downloads/brochures/Proxim_small_call_backhaul.pdf, 2013.

[252] K. Doppler, M. Rinne, C. Wijting, C. Ribeiro, and K. Hugl, "Device-to-device communication as an underlay to LTE-advanced networks," *IEEE Commun. Mag.*, vol. 27, no. 12, pp. 42–49, Dec. 2009.

[253] P. Janis, C. Yu, K. Doppler, C. Ribeiro, C. Wijting, K. Hugl, O. Tirkkonen, and V. Koivunen, "Device-to-device communication underlaying cellular communications systems," *Int. J. Commun. Netw. Syst. Sci.*, vol. 2, no. 3, pp. 169–178, 2009.

[254] S. Hakola, T. Chen, J. Lehtomäki, and T. Koskela, "Device-to-device (D2D) communication in cellular network - performance analysis of optimum and practical communication mode selection," *Proceedings of IEEE Wireless Communications and Networking Conference (WCNC'2010)*, pp. 1–6, April 2010.

[255] CISCO Systems, "Cisco service provider Wi-Fi: A platform for business innovation and revenue generation," *Solution Overview*, pp. 1–12, June 2012.

[256] H. Tabassum, M. Z. Shakir, and M. Alouini, "Area green efficiency (AGE) of two tier heterogeneous cellular networks," *Proceedings of IEEE Conference on Global Communications, (GLOBECOM'2012)*, pp. 529–534, Dec. 2012.

[257] S. Tombaz, P. Monti, W. Kun, A. Vastberg, M. Forzati, and J. Zander, "Impact of backhauling power consumption on the deployment of heterogeneous mobile networks," *Proceedings of Conference on Global Communications, (GLOBECOM'2011)*, pp. 1–5, Dec. 2011.

[258] J. Baliga, K. Hinton, and R. S. Tucker, "Energy consumption of the Internet," *Proceedings of Joint International Conference on Optical Internet, 2007 and the 32nd Australian Conference on Optical Fibre Technology (COIN-ACOFT'07)*, pp. 1–3, June 2007.

[259] B. Skubic and D. H. Ericsson, "Evaluation of ONU power saving modes for gigabit-capable passive optical networks," *IEEE Mag. Network*, vol. 25, no. 2, pp. 20–24, 2011.

[260] S. Verdu, "Spectral efficiency in the wideband regime," *IEEE Trans. Inf. Theory*, vol. 48, no. 6, pp. 1319–1343, Jun. 2002.

[261] G. Auer *et al.*, "How much energy is needed to run a wireless network?" *IEEE Wireless Commun. Mag.*, vol. 18, no. 5, pp. 40–49, Oct. 2011.

[262] Wolfram|Alpha, http://reference.wolfram.com/language/ref/Gamma.html, Online accessed, Sept. 2014.

[263] A. Cuyt, V. B. Petersen, B. Verdonk, H. Waadeland, and W. B. Jones, *Handbook of Continued Fractions for Special Functions*, Springer-Verlag, 2008.

[264] NSN, Nokia solutions and networks looking ahead to 5G, White paper, Nokia Solutions and Networks Oy, Finland, Dec. 2013.

[265] A. Radwan and J. Rodriguez, *Energy Efficient Smart Phones for 5G Networks*, Springer-Verlag, 2015.

[266] A. Asadi, Q. Wang, and V. Mancuso, "A survey on device-to-device communication in cellular networks," *IEEE Commun. Surv. Tutorials*, vol. 16, no. 4, pp. 1801–1819, Nov. 2014.

[267] Q. Wang and B. Rengarajan, "Recouping opportunistic gain in dense base station layouts through energy-aware user cooperation," *Proceedings of IEEE International Symposium on World of Wireless, Mobile and Multimedia Networks, (WoWMoM)*, pp. 1–9, 2013.

[268] L. B. Le, "Fair resource allocation for device-to-device communications in wireless cellular networks," *Proceedings of IEEE Conference on Telecommunications, (GLOBECOM)*, pp. 5451–5456, 2012.

[269] T. Han, R. Yin, Y. Xu, and G. Yu, "Uplink channel reusing selection optimization for device-to-device communication underlaying cellular networks," *Proceedings of IEEE International Symposium on Personal Indoor, and Mobile Radio Communications, (PIMRC)*, 2012.

[270] C. H. Yu and O. Tirkkonen, "Device-to-device underlay cellular network based on rate splitting," *Proceedings of IEEE Conference on Wireless Communications and Networking, (WCNC)*, 2012.

[271] X. Chen, L. Chen, M. Zeng, X. Zhang, and D. Yang, "Downlink resource allocation for device-to-device communication underlaying cellular networks," *Proceedings of IEEE International Symposium on Personal Indoor, and Mobile Radio Communications, (PIMRC)*, 2012.

[272] M. J. Yang, S. Y. Lim, H. J. Park, and N. H. Park, "Solving the data overload: device-to-device bearer control architecture for cellular data offloading," *IEEE Veh. Technol. Mag.*, vol. 8, no. 1, pp. 31–39, 2013.

[273] C. Isheden and G. P. Fettweis, "Energy-efficient multi-carrier link adaptation with sum rate-dependent circuit power," *Proceedings of IEEE International Conference on Global Communications, (GLOBECOM)*, pp. 1–6, Dec. 2010.

[274] L. M. Ausubel and P. Milgrom, *Chapter 3: Ascending Proxy Auctions - Combinatorial Auctions*, MIT Press, 2015.

[275] L. Lei, Y. Zhang, X. Shen, C. Lin, and Z. Zhong, "Performance analysis of device-to-device communications with dynamic interference using stochastic Petri nets," *IEEE Trans. Wireless Commun.*, vol. 12, no. 12, pp. 6121–6141, Dec. 2013.

[276] M. Ali, J. M. Zain, M. F. Zolkipli, and G. Badshah, "Mobile cloud computing & mobile battery augmentation techniques: a survey," *Proceedings of IEEE Student Conference on Research and Development (SCOReD)*, pp. 1–6, Dec. 2014.

[277] J. D. Power Associate, Wireless Smartphone Customer Satisfaction, vol. 1, http://www.jdpower.com/sites/ default/files/2012030-whst.pdf, Mar. 2012.

[278] DigiKey Electronics, Battery Life Calculator, http://www.digikey.com/en/resources/conversion-calculators/conversion-calculator-battery-life, Online accessed, Jul. 2015.

Index

Page numbers in *italics* indicate figures; page numbers in **bold** indicate tables